SAUNDERS COMPLETE PACKAGE
FOR TEACHING GENERAL PHYSICS

Melissinos and Lobkowicz: *Physics for Scientists and Engineers* — Volume 1

Lobkowicz and Melissinos: *Physics for Scientists and Engineers* — Volume 2

Slides to Accompany *Physics for Scientists and Engineers* — Volumes 1 and 2

Greenberg: *Discoveries in Physics for Scientists and Engineers, A Laboratory Approach* — Second Edition

Serway: *Concepts, Problems and Solutions in General Physics* — Volumes 1 and 2

Davidson and Marion: *Mathematical Preparation for General Physics with Calculus*

discoveries in physics

for scientists and engineers

LEONARD H. GREENBERG

University of Saskatchewan, Regina Campus,
Regina, Canada

SECOND EDITION

 SAUNDERS GOLDEN SERIES

1975

W. B. SAUNDERS COMPANY

PHILADELPHIA · LONDON · TORONTO

W. B. Saunders Company: West Washington Square
Philadelphia, PA 19105

12 Dyott Street
London, WC1A 1DB

833 Oxford Street
Toronto, Ontario M8Z 5T9, Canada

Library of Congress Cataloging in Publication Data

Greenberg, Leonard H
  Discoveries in physics for scientists and engineers.

  (Saunders golden series)
  First ed. published in 1968 under title: Discovery in physics.
  Includes index.
  1. Physics—Experiments.   I. Title.
QC33.G68   1975   530'.028   74-4565
ISBN 0-7216-4246-2

Discoveries in Physics for Scientists and Engineers   ISBN 0-7216-4246-2

Last digit is the print number: 9   8   7   6   5   4   3   2   1

PREFACE TO SECOND EDITION

In preparing the second edition of this book, I have taken into account the comments from those who used the first edition. Not only have these led to some additions, corrections and changes, but also it was apparent that some further remarks would be advisable concerning the function of the laboratory and how this book approaches it. In general, the organization of the first edition has been maintained.

The almost universal changing attitude of students toward the university has also been considered. Students have been challenging the offerings of the university. They want either to run it their way or, alternatively, to know exactly why they should do what they are asked to do. Rather than changing many of the experiments for which the value may not have been clear to the student, the reasons for their inclusion have been inserted. The reason may be to let the student know that a particular project is an exercise in precision, that precise equipment is being used to do a good job, or that the phenomenon being worked with is one that will probably be of value later if the student continues to study science. But the idea is that if the student knows why he is spending his time on a certain project, he will then do it well.

The laboratory is a place of learning, and the lab instructors are of paramount importance. In conducting a laboratory session the instructor should present the aims of the project clearly, so that the students understand the goal and, through explanations of the objectives, will be stimulated to carry out the projects. The operations of the particular apparatus to be used in the work should then be outlined. If the students know what to find and can handle the equipment, they need little additional help. Above all, they must be challenged to find out something in the laboratory. The enthusiasm of the student can be lost if the answer is known beforehand, for to try to learn what you already know is a futile exercise. For this reason a prior theoretical analysis is not advised, and many of the laboratory projects should precede the lectures on the subject.

The level of the experiments varies, but this is to allow the instructor to select experiments of an appropriate level. Originally, a chart showing the level of each experiment was to be included, but this was discarded because the students should not be made to feel that they are at a low level (if they are). The instructors can choose the experiments, and the enthusiasm can be the same, whatever the level.

Some new experiments have been added, one of which is the measurement of the resolving power of a telescope. This is an introduction to the phenomenon of resolution, which is encountered in many areas of physics. Included in the project is a measurement of resolution of the eye; this is a very brief introduction to an aspect of biophysics. Resolution is a good lab project, because theoretical treatments are usually based on the Rayleigh criterion, which is based on an arbitrary limit. In the project outlined the results usually slightly exceed those set by the theoretical treatment.

A photographic method for measuring spectra has been included. The principal advantage over the spectrometer lies in the precision. The method allows wavelength measurement accurate to a nanometer, although students are often off by 2 or 3. Yet the precision is such that element identification is definite. The project can also be extended easily to stellar spectra; this provides an introduction to astrophysics.

A project on measurement of speed distribution of thermally emitted electrons has also been added. This is inclined to be a "black box" type of experiment, but an attempt was made to show how the ideas are arrived at theoretically and that the visible meter readings are what one would expect. The student learns to work with, and understand, the invisible. The energy distribution of electrons around a hot cathode is Maxwellian, and the results can be converted directly to speeds of gas molecules. High speed aircraft problems and the space re-entry problem are then demonstrated.

Concerning major changes in the experiments that were also in the first edition, the project on apparent depth has been rewritten to indicate its relevance in many areas of science. It is a simple but challenging project; since it will probably be one of the first, difficulty will still be encountered in reading the microscope verniers. The physics to be learned is not profound, although it can be important. Another elementary experiment, analysis by measurement of density, has been changed to have more relevance to physics.

In general, the majority of changes from the first edition are small and designed to make the lab program more inspiring. It must be emphasized, however, that the ultimate responsibility for a good laboratory experience lies with the instructors.

I am grateful to those who have submitted criticisms of aspects of the first edition, although not all have been acted upon. For example, it was suggested that since the unbalanced bridge was described in the text, there should be a project using it. One idea is to construct a thermometer, perhaps using a tungsten filament from a light bulb (break away the glass). Another idea, which could be suggested to some capable students, is to use the bridge that is given for the measurement of precision of resistors as an unbalanced bridge. Given only the idea, a good student can set it up, calibrate it and use it.

The chapter on equipment, Chapter 10, which was originally envisaged as an Appendix, has been left at the end of the book even though reference is made to it in many of the projects.

Some material on the camera has been inserted in Chapter 10. A camera is specified for use in many of the experiments, but in very few physics books is any useful material on the camera included. Because of this, more information than is needed for the projects has been included. In particular, the relationships that allow the camera to be used as a measuring instrument are discussed. The exposure systems, focal ratio and Exposure Value are both described. Such material may be of benefit to many students.

In conclusion I would like to stress that in conducting the student laboratory an attempt should be made to duplicate the research laboratory attitude. The answer to a problem is to be found in the lab, and remember too that research scientists work best on projects when they become excited enough to find the answer. The same is true for students, so try to encourage them to want to find the answers.

There are many to whom I would like to express thanks for their assistance in preparing this edition. Foremost is Dr. F. Lobkowicz of the University of Rochester, whose extensive comments were greatly appreciated. Included also are the many other critics of the first edition, the students, fellow faculty, the staff at W. B. Saunders Company, and also Miss Sheilla Fruman and Miss L. Grönberg, who labored diligently with the typing and the correction of my many errors.

L. H. Greenberg

CONTENTS

INTRODUCTION

It does not very much matter to science what these vague speculations lead to, for meanwhile it forges ahead in a hundred directions, in its own precise experimental way of observation, widening the bounds of the charted region of knowledge, and changing human life in the process. It may be on the verge of discovering vital mysteries, and yet they may elude it. Still it will go on along its appointed path, for there is no end to its journeying. Ignoring for the moment the Why? of philosophy, it will go on asking How?, and as it finds this out it gives greater content and meaning to life, and perhaps takes us some way to answering the Why.[1]

In this way did a modern statesman so beautifully describe the scientific method and the scientific purpose.

Of the people called scientists there are those who speculate and theorize, those who observe and measure, and those who fit science into a philosophy of life.

The physicists who make it their chief occupation to observe the world are classed as experimental physicists. In this category have been such great men as Michael Faraday (electricity), the Curies (radioactivity), and Lord Rutherford (nuclear physics). Their counterparts, the theoretical physicists, are those who use reasoning, primarily mathematical, to form theories which explain the phenomena observed and predict new phenomena. Galileo and Newton, who worked at the beginning of this scientific era, were of necessity both theoretical and experimental physicists. In more recent times men such as Henri Poincaré and Albert Einstein have been outstanding examples of theoretical physicists. But no system of categorization can be rigid and in most instances there is a combination of knowledge and ability in both areas.

The two aspects of physics have been inseparable for the progress of the subject, and they must not be separated at the level of introductory physics in a university. If one is to understand the importance and the power of experimental physics, his laboratory experiences must show the actual ways in which experiments complement theoretical work. Physical phenomena will be observed, measurements made, and the data analyzed to yield some type of information: a physical constant, a choice between theories, or perhaps a description of a relationship between two quantities which may amount to the formulation of a physical law. The object of an experiment should never be to verify a theory or a law. The use of the word "verify" indicates that the theory is already accepted as being correct. If an experimenter sets out to verify a law, he will be biased, and bias does not go hand in hand with good experimental work.

One reason why a law can never be shown to be absolutely true, which is what is implied by *verified,* is that no measurements are exact; and no conclusion drawn on the basis of measured values can be accurate to a precision higher than that of the measurements. Consequently, no law in physics is regarded as being exact. The law of the conservation of energy (including mass) is one of the basic laws of physics, and it has been shown to hold up to at least one part in 10^{12} or a million million.

It may seem to be enough to say that the law of conservation of energy is not exact, but it is important to know the limit, as the following example shows. Observations made by astronomers show all the galaxies receding from each other. It seems that as time goes on the average density of matter in large volumes of space decreases. This has led to the "expanding universe" and the "big bang" models of the universe. Another observation has been that in any direction at a given distance there seems to be the same number of galaxies. The universe seems everywhere the same, on the average, and the theory has been put forward that we have no unique position in space and probably also no unique position in time; this is the "steady state" model of the universe. However, the apparent expansion must be accounted for, and this can be done by assuming that matter is slowly created and gradually collects to form new galaxies and maintain a constant average spacing.[2] This may seem to violate the concept of the conservation of energy, but a calculation shows that all that is necessary to allow such creation of matter is that the law be violated to one part in 10^{15}. If the law can be shown to be that precise, the steady state model of the universe with its continuous creation concept must be discarded.

The conservation of mass alone used to be taught as an inviolable law, but sufficiently precise measurements would have shown that, as we now know, mass is a form of energy and the conservation law includes both. Thus, new things are learned when the limits of the old laws are found.

The measurement of constants is one of the purposes of experimental physics. The work in the vast physics research laboratories which dot the world is not devoted entirely to this aspect of the subject; in fact, to this function is devoted an exceedingly small fraction of their effort. More important, the introductory laboratory experience will demonstrate how physics probes the unknown, how experiment can guide theory, and how theory can guide experiment.

It is not necessary that you perform in the laboratory what has previously been explained in class; indeed, it is often necessary that the laboratory work precede the lectures so that you are able to see how a law can be obtained in the laboratory. Then the fact that the theory suggests the same law which was obtained experimentally will confirm the usefulness of the theory and will give you faith in both the experimental method and the theoretical analysis. Therefore, there are experiments and problems designed to illustrate methods of analyzing data to find relations between quantities, for physical laws are nothing more than the description of such relations. Snell's law describes how the angle of incidence of a light ray striking a boundary between two materials is related to the angle of refraction of the ray after it has passed into the second medium; Boyle's law describes the relation between the pressure and the volume of a given mass of gas at a constant temperature. Once the general methods of analysis which are described have been mastered, they may be used to analyze almost any given set of data to find the relation between the quantities represented. Even the non-science student, who perhaps may not fully master the methods, will gain an understanding of one of the functions of experimental physics.

It is also one function of experimental work to determine constants. The design of an experiment to determine a constant is basically different from that of an experiment designed to find a relation between quantities. This difference is described in the book, and some of the experiments outlined are designed for the purpose of determining constants. One basic difference is that in the determination of a constant some physical law or laws are assumed to be correct and certain constants in the laws are determined.

Two of the other aspects of experimental work are common to both types of experiment. These are the handling and use of apparatus and the understanding and calculation of experimental error. A portion of the book is devoted to the description of various common types of measuring instruments, and one chapter discusses experimental error. Some of the experiments require a knowledge of how precise the answer is before a conclusion can be satisfactorily drawn. The material describing the calculation of errors is of importance, therefore, throughout much of the laboratory work.

There are many other purposes of a physics laboratory period. Some of these are the familiarization with phenomena which are not part of everyday experience, the appreciation of the magnitude of various quantities, and the introduction to the use of apparatus—although for some the detailed working is not understood but the function may be simply learned. In this last category are such devices as the stroboscope, the Geiger-Müller counter and its scaling circuit, and various timing devices.

The experiments will require the repeated use of certain pieces of measuring equipment with which you will become familiar and facile. Then the emphasis can be placed on the physics of the experiment, rather than on puzzling out a new piece of measuring apparatus. A measuring microscope, for example, will be used in many experiments. It is also suggested that a multimeter be used as much as possible in electrical measurements so that you will become familiar with this common instrument.

The proper recording of measurements at the time the experiment is performed is emphasized and the two forms of laboratory reporting, "lab record" and "formal report," are described. Throughout the book frequent references are made to original works and to other texts which have been chosen partly because they would be good supplementary reading for students at this level. Reference to authoritative or original work is one of the characteristics of laboratory reporting. The importance of going to the original source will be illustrated and your education will certainly be enriched by the investigation of the reference material.

It is hoped that the laboratory exercises will be inspiring to you because they present the methods and analysis used in research, because you will not usually know the answer expected, and because you will become familiar with some of the modern pieces of equipment used in these experiments.

An experiment will usually involve many aspects of laboratory work, such as the operation of measuring devices, the analysis of data, and the evaluation of errors. Thus, you will probably have to refer to several parts of the book in each experiment. If you are confronted with a vernier caliper, look in the index and you will find that in Chapter Ten the reading of a vernier scale is described, as are many other pieces of measuring apparatus. To find how to analyze data by using a graph, refer to the appropriate part of Chapter Five. When you come to work out the precision of your result, find the pertinent portion of Chapter Three, and for the writing of reports read Chapter Four. In general, because the main part of each experiment deals with the material of the chapter which includes that experiment, it is a good thing to have read that chapter before doing the experiment. It would even be advisable to familiarize yourself with the content of the whole book early in the term.

Before you do any of the experiments get a notebook—coil back or bound in some way, but preferably not the loose-leaf style; the loose-leaf pages are too easily taken out and destroyed. Use the notebook to record all the measurements you make along with what is measured in each case. This will be your lab record book. Make it as messy or as neat as you want to. You will find, however, that a degree of neatness helps you to avoid confusion and error. The notebook will contain your wrong measurements and your right ones—even your spur-of-the-moment suggestions for better ways of doing things. It is from this lab record that the report sheets will be made, but it will not be graded except insofar as it is being used for recording. It will teach you one of the maxims of research: *record all data directly into a permanent record book.*

Date your work so that you can refer back to it, and put words with each measurement so that you will know, on reference, what the numbers are about.

Every scientist needs such a record, and you are becoming familiar with the ways of the scientist!

REFERENCES

1. Nehru, J.: The Discovery of India. New York, Doubleday and Company, Inc., 1954, p. 19.
2. Bondi, H., Bonnor, W. B., Lyttleton, R. A., and Whitrow, G. J.: Rival Theories of Cosmology. London, Oxford University Press, 1960, pp. 17, 42.

1

ARISTOTLE, GALILEO, AND EXPERIMENTAL SCIENCE

Aristotle is widely known for saying that objects fall at speeds proportional to their weights, and Galileo for showing that objects fall at the same speed. Aristotle, it is often said, performed no experiments and Galileo originated experimental science. These two great men both actually observed objects falling under the action of gravity, but they came to very different conclusions. Why? Was one wrong and the other right? What were the differences in their reasoning? An analysis of their observations and of their methods of reasoning shows that there was an element of truth in the conclusions of both men.

Aristotle, who possessed one of the great minds of all time, observed that objects fell at different rates. In general, the light objects fell more slowly than the heavy ones and he drew conclusions from his observations. According to Galileo,[1] Aristotle expressed it thus: ". . . bodies of different weights, in the same medium, travel (in so far as their motion depends upon gravity) with speeds which are proportional to their weights." Did Aristotle verify this experimentally? In his writings he apparently refers to having done the experiment, since he uses the expression "we see."[2] Any of us can see the same thing by performing a simple experiment. Drop a small piece of chalk and compare its rate of fall with that of some particles of chalk dust. Does this not show that the heavier object falls much faster than the lighter object? Aristotle implied, however, that "two stones, one weighing ten times as much as the other, if allowed to fall, at the same instant, from a height of, say, 100 cubits, would so differ in speed that when the heavier had reached the ground, the other would not have fallen more than 10 cubits."[2] An even more striking example: "An iron ball of one hundred pounds falling from a height of one hundred cubits reaches the ground before a one-pound ball has fallen a single cubit."[3] Galileo counters with "You find, on making the experiment, that the larger outstrips the smaller by two finger-breadths, that is, when the larger has reached the ground, the other is short of it by two finger-breadths; now you would not hide behind these two fingers the ninety-nine cubits of Aristotle."[3]

As a result of this and other observations with balls of different density Galileo announced, ". . . I came to the conclusion that in a medium totally devoid of resistance all bodies would fall with the same speed."[4] Galileo did not make the general, unqualified statement that all objects fall with the same speed.

Aristotle was not referring to the motion of objects in a medium totally devoid of resistance, but referred always to motion in a resisting medium. The results of Galileo and of Aristotle are thus not exactly comparable because they refer to different conditions. We may consider that Aristotle, in observing small falling objects, was in reality observing their terminal velocity. When an object falls from rest in a resisting medium, as the speed increases so does the resistance force. The speed increases until it eventually reaches a limiting value at which the resistance force balances the weight. This limiting value is called the terminal velocity. This condition certainly describes the case of powdered rock or chalk. Furthermore, it is true that the terminal velocities of objects falling in a viscous medium are proportional to their weight in that medium if they are of the same external size and shape and if the conditions are such that the resistance force is proportional to the velocity. The expression "proportional to their weight in that medium" is used here to mean the difference between the gravitational force and the buoyant force. The buoyant force results from the action of gravity, so that the condition mentioned by Aristotle (in so far as the motion of the objects depends upon gravity) still holds, and Aristotle is then at least partially vindicated.

We cannot say that one of these men was right and the other was wrong; neither was the work of either of them completely above reproach. Galileo was (and I might add that many authors of physics texts since his time also have been) unduly critical of the work of Aristotle. We can, instead of condemning, learn some important facts about experimental work from these two great men.

The necessity for measurement, not just observation, is the first lesson. It is possible to observe relations without measuring them; but unless measurements are made, the relation between any two quantities cannot be accurately described. Had Aristotle made some measurements, he would possibly have found that there was not a direct relation between velocity and weight for the experiments that he was able to carry out.

Incidentally, a direct relation implies that one quantity varies as the first power of the other. If y varies directly as x, then y and x are related by a formula of the type $y = kx$. Only after experimental measurements of two quantities have been analyzed to see whether this type of relation describes the situation is it possible to make the statement that "one quantity varies directly as the other." In some situations it is found that the dependence is of some other form such as an inverse relation, an inverse square, or indeed, any one of a large number of possibilities. One of the services of experimental physics is to find the relation between two quantities under some set of physical conditions.

Not always was it known that the force of attraction between two charged particles varies inversely as the square of the distance. Cavendish in 1771 wrote in a memoir to the Royal Society, "There is a substance which I call the electric fluid, the particles of which repel each other and attract the particles of all other matter, with a force inversely as some less distance than the cube."[5] The determination of an unknown relation is always more difficult than supposedly "verifying" one that has been already found or that is assumed.

The many inverse square laws which are now generally accepted have been found by experiment to describe various phenomena with great precision. The inverse square law is, in fact, often used to calibrate equipment. A photocell (or "electric eye") can be used to measure light intensity by measuring the current produced by the cell when light falls on it. The question arises, If the light intensity is doubled, will the current be doubled? In other words, is the current output proportional to the light intensity? To see whether it is, the photocell can be mounted near a concentrated light source and the current output determined. (In doing this, care must be taken to avoid light reflection from such things as the table top, shiny apparatus, or light colored clothing.) Then it can be moved to another position and the new current output determined. The change in intensity of light can be calculated from the measured distances using the inverse square law and then compared with the change in current output of the cell. It will often be found that the current output is not quite proportional to the light intensity and photocells must be calibrated in order to convert from current reading to light intensity.

In practice there is frequently a deviation from the inverse square law. In the case of light intensity the size of the source must be small compared with the distance to the detector or there will be a deviation from the law. For a gravitational field to be of an inverse square form, the central mass must be spherically symmetric. There is some doubt at the moment whether the unusual form of the orbit of the planet Mercury (see Fig. 1-1) is due to relativistic effects (see p. 5), or whether it is due to a nonspherical distribution of mass of the sun, perhaps to a slight bulge at its equator. Artificial planets may help us to solve this problem.

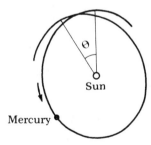

Figure 1-1 The orbit of Mercury is a rotating ellipse.

The problem of determining relations is still with us at the forefront of research. One of the many problems today is the effect of radiation on material. After copper, for example, has been placed in a nuclear reactor and irradiated with neutrons, a larger stress is required before reaching the elastic limit than if it had not been irradiated. The mechanism of the irradiation effect is apparently not understood, for in reporting the result of some of his experiments, Dr. D. S. Billington says, "Yield stress increase induced by irradiation of copper varies as

[the] cube root of [the] fast neutron dose, while most theories indicate a square-root dependence. Lack of [a] suitable explanation seems [the] chief hurdle to understanding [the] mechanism of radiation-induced hardening of metals."[6] Later work has confirmed these results but theoretical interpretation has not yet been achieved.

Without measurement, the observation would be merely that there was a relation between the yield stress of the copper and the radiation dose to which it had been exposed. As the dose increased, so did the yield stress of the copper. According to theory, a square root dependence was expected, and only an experiment could determine what the exponent in the relationship actually was and thus show whether or not the theory was adequate.

So important is the inclusion of measurements with the observation that in modern experimental science the word *observation* implies measurement, though in everyday use only the act of perceiving is understood. Funk and Wagnall distinguish between the nonscientific and the scientific definitions of the word in this way:

Observation:

1. the act, power, or habit of observing or taking notice; the act of perceiving or of fixing the powers of sense or intellect or anything. . . .

2. scientific scrutiny of a natural phenomenon, especially a visible fact or phenomenon, for experiment, verification, or measurement and calculation. . . .

There is still another thing that we can learn from the work of Galileo and Aristotle. We can say that their statements refer to different conditions. Aristotle spoke of motion in a resisting medium; Galileo extrapolated his results to probable motion in a vacuum. Galileo spoke of a situation that he was actually not able to achieve experimentally. Keep in mind that he, too, observed that objects fell at different speeds in air. Each time he describes an experiment he says that the heaviest object fell the fastest; but by experimenting with a series of objects on which the relative effect of air resistance successively diminished, he saw that if air resistance could be completely eliminated, then all objects would probably fall at the same rate. This was actually a bold assumption. He did not have measurements to verify the statement, but it was later shown to be correct. Had it been shown to be incorrect, Galileo would not be revered to the extent that he now is. Much scientific progress is the result of just such bold, simple generalizations.

This common version of the history of the understanding of falling objects contains apparently only part of the truth. Aristotle reasoned that in a vacuum all objects would fall at the same speed, but he considered a vacuum to be impossible. Lucretius, who lived almost 200 years after Aristotle, wrote:

> When objects fall straight down through water or thin air,
> The heavier faster falls, just because the medium
> Checks not all things in equal measure equally,
> But rather faster yields to heavier things.
> And empty space can never, anywhere, check things in flight.
> Its nature makes it yield.
> And so both heavy things and light
> Are borne at equal pace through silent void.[7]

Galileo seems to have resurrected this long-lost conclusion described so well by Lucretius 1700 years earlier. Galileo did contribute the concept that in a vacuum all objects would fall with the same constant *acceleration*. That is, he described the motion in a precise way.

In making a statement, we see the necessity of specifying the conditions. Beginning with the work of Galileo and Newton, a mechanics was formulated to describe the motion of objects under the action of forces. This is known commonly as Newtonian mechanics, and describes the motion of bodies of the solar system to a very high degree of accuracy, but when it is applied to the orbit of Mercury, the actual motion is not that predicted by Newtonian mechanics. The axis of the ellipse describing the orbit rotates, as illustrated (although exaggerated) in Figure 1-1, by 43″ per century more than can be accounted for by the attraction of the other planets. Newtonian mechanics seems to be not completely successful in this case, but just that much rotation of the axis of the ellipse is predicted by the general relativity theory.

Newtonian mechanics describes the motion of atoms and electrons, provided that the particle velocity is much less than the velocity of light. Relativistic mechanics, based on Einstein's work, gives a more accurate description of motion at velocities approaching the velocity of light. We do not say that Newtonian mechanics is wrong; we say that it describes motion at speeds which are small compared with the speed of light and that at high speeds, relativistic mechanics must be used.

Galileo and the Pendulum

Any clock or watch is controlled by some recurrent phenomenon, and the precision depends on the constancy of that phenomenon. For hundreds of years the most precise clocks were controlled by a swinging pendulum. Oscillating balance wheels, vibrating tuning forks, and quartz crystals are examples of devices used to regulate clocks today. Before a new phenomenon is used to control a timing device, the factors that affect the period must be found so that they can be reduced to a minimum.

It was Galileo who first noticed the constancy of the period of the swing of a pendulum and adapted it to control a clock, although Christian Huygens is credited with the actual development of the clock. In this project you, like Galileo, are to investigate one aspect of the pendulum for clock control, that is, the effect of variation in amplitude.

The following statements regarding the period of a simple pendulum are taken from the translation of a book by Galileo Galilei called *Dialogues Concerning Two New Sciences.* The pages refer to the Dover publication first presented in 1914.[1] The book first appeared in English in 1665.

> ...I never dreamed of learning that one and the same body when suspended from a string a hundred cubits long and pulled aside through an arc of 90 degrees or even 1 degree or 1/2 degree, would employ the same time in passing through the largest of these arcs.
>
> ...each pendulum has its own time of vibration so definite and determinate that it is not possible to make it move with any other period (altro periodo) than that which nature has given it [p. 97].

On page 254, again talking about the simple pendulum he says:

> ...if two persons start to count the vibrations, the one the large, the other the small, they will discover that after counting tens and even hundreds they will not differ by a single vibration, not even by a fraction of one.

From this Galileo makes the following proposition:

> This observation justifies the...following proposition...namely, that vibrations of very large and very small amplitude all occupy the same time [p. 255].

The object of this experiment is to check Galileo's statements by making measurements of the period of a simple pendulum at several different amplitudes.

The following are hints and instructions which will enable the student to make the measurements with enough precision to check Galileo's statements:

1. The period of oscillation is the time between two successive passages past the same point and going in the same direction.

2. The apparatus must be clamped securely to the desk in order to preserve stability when the pendulum is moved through large amplitudes. Further, the pendulum support must be tightly secured so that no variation in point of support, and hence of length, occurs.

3. The amplitude of oscillation is the angle θ, through which the pendulum swings on either side of the rest position.

4. To determine the period, find the time for at least 25 oscillations; then calculate the time for one oscillation. The lowest point is the best reference because the pendulum passes more quickly through that point than any other, and hence the time at which it passes can be most accurately determined.

5. Find the period for the full range of amplitudes and record your results in a table. It will probably be necessary to make several measurements at each amplitude in order to come to a definite conclusion.

Apparatus

A solid support for the pendulum — with apparatus for reading amplitude directly

1 meter of fishline or other nontwisted cord

A drilled metal ball about 1 inch in diameter

A stop watch reading to tenths of seconds or, preferably, to hundredths

A protractor

REFERENCES

1. Galilei, G.: Dialogues Concerning Two New Sciences. New York. Dover Publications, 1914, p. 65.
2. Ibid., p. 62.
3. Ibid., p. 64.
4. Ibid., p. 72.
5. Turner, D. M.: Makers of Science — Electricity and Magnetism. London, Oxford University Press, 1927, p. 32.
6. Billington, D. S.: Relaxing reliance on empirical data. Nucleonics. 18:64 (1960).
7. Winspear, A. D.: Lucretius and Scientific Thought. Montreal, Harvest House, 1963, p. 107.

2

THE LAWS OF
PHYSICS

About us is a world in continuous motion. People move; cars move; plants grow; sound waves crisscross; light waves are everywhere; airplanes drone overhead; and at night even the stars move across the sky. The scientist looks for an order in this seeming chaos. Order must be sought in just one small area at a time, and it is because of this that the scientist often takes into his laboratory a small fragment of the external world for investigation. Perhaps, before the time of laboratories, order was first obvious in the motion of the sun, moon, and stars; hence, astronomy became the first science. Each day the sun moved across the sky. The stars moved in orderly procession across the sky each night. The planets moved in what seemed at first to be erratic paths among the stars, but eventually it became possible to predict the motion of even these wanderers. Eclipses of the sun and moon occurred without warning, but, as data accumulated, they were seen to occur in a pattern repeating itself over a basic period lasting approximately 18 years. Order replaced chaos as the science of astronomy developed.

Before the advent of science, each event that occurred was considered to be under the direct order of one or more of a kind of super being. Poincaré said:

> You know what man was on the earth some thousands of years ago, and what he is today. Isolated amidst a nature where everything was a mystery to him, terrified at each unexpected manifestation of incomprehensible forces, he was incapable of seeing in the conduct of the universe anything but caprice; he attributed all phenomena to the action of a multitude of little genii, fantastic and exacting, and to act on the world he sought to conciliate them by means analogous to those employed to gain the good graces of minister or a deputy. . . .
>
> To-day we no longer beg of nature; we command her, because we have discovered certain of her secrets and shall discover others each day. We command her in the name of laws she can not challenge because they are hers; these laws we do not madly ask her to change, we are the first to submit to them. Nature can only be governed by obeying her.[1]

The discovery of laws or, in other words, the finding and description of an order in nature is one of the principal functions of science. How do scientists go about finding laws? The answer to that question would be a description of a process of science. A law is not the result of idle speculation or random guessing. The first step in discovering a law is the gathering of data about some phenomenon. The data are then analyzed in an attempt to find some correlation between quantities; and the description of this relation, usually in mathematical terms, amounts to a law. Sometimes a law is no more than a mathematical formula; and sometimes formulae, though not usually called laws, could be referred to as such.

Analysis of the experimental data will not show a law exactly, but the law will be strongly suggested by the experiment. After several laws concerning a phenomenon have been found, theories or models can be set up to describe the working of the phenomenon. A theoretical analysis can then be carried out and more laws may be suggested.

These laws should then be looked for experimentally. We prefer not to say "verified." The analysis of the experimental data should be to see what the relation or law really is, and if the experimental law is sufficiently close to that predicted by the theory, then it will have been shown that the theory or model is a useful one. The experiment does not show that the model is correct, because the predictions of the model may not agree with those of some other experiment at another time. If that happens, the theory or model will be shown to be limited in its usefulness. Subsequently, it may either be used in its limited area or it may be completely replaced by another theory or model which adequately describes a larger range of situations.

Furthermore, each law will be valid only under certain conditions or limiting factors. If we know the limiting conditions, we know more about the law than if we do not. A scientist should always feel uneasy about a law until he has found its limitations, for experiments are showing limitations to more and more of our laws and we can, in fact, assume that all laws have limits whether we have found them or not.

The search for limitations on the laws is described by Poincaré in this manner:

> So when a rule is established we should first seek the cause where this rule has the greatest chance of failing. Thence, among other reasons, come the interest of astronomic facts and the interest of the geologic past; by going very far away in space or very far away in time, we may find our usual rules entirely overturned, and these grand overturnings aid us the better to see or the better to understand the little changes which may happen nearer to us, in the little corner of the world where we are called to live and act. We shall better know this corner for having travelled in distant countries with which we have nothing to do.[2]

Sir Isaac Newton recognized the limitations of physical laws when he wrote in *Opticks:*

> And although the arguing from Experiments and Observations by Induction be no Demonstration of general Conclusions; yet it is the best way of arguing which the Nature of Things admits of, and may be looked upon as so much the stronger, by how much the Induction is more general. And if no Exception occur from Phenomena, the Conclusion may be pronounced generally. But if at any time

afterwards any Exception shall occur from Experiments it may then begin to be pronounced with such exceptions as occur.[3]

D. H. Wilkinson, speaking at Oxford in 1958, recognized this too, for he said,

> ... it must be emphasized that any conservation law is experimental in origin and cannot be asserted more firmly than its experimental verification warrants.[4]

Henri Poincaré recognized it when he said,

> ... no particular law will ever be more than approximate and probable[5] ... the conservation of electricity ... has been found exact, at least until now.[6]

The important words here are "at least until now," for they show that in his mind he accepts the idea that some day limitations may be defined for that law.

In the translator's introduction to Poincaré's *The Value of Science,* the original edition of which was published in 1912, there is the following statement:

> In the present book we see the very foundation rocks of science, the conservation of energy and the indestructibility of matter, beating against the bars of their cages.[7]

History now shows that under certain conditions it is necessary to modify these concepts. Mass is a form of energy, and when we speak of the conservation of energy, we also mean mass. No laws fell—limitations were put on the existing ones. In much of physics we can still speak of the conservation of energy and of the conservation of mass separately. There is, however, a new class of phenomena which require the new law for their description.

Putting limits on a law is like putting up "no smoking" signs in some rooms and signs which read "smoking allowed" in other rooms. Then, in any particular room, we know exactly what we may or may not do. If we know the limits to a law, we know in what situations it will give a satisfactory answer and in what situations it will not.

A simple form of a law may describe a situation or phenomenon to a degree of accuracy which in some cases is sufficient but in others is not. For instance, the relation between the pressure, P, and the volume, V, of a given mass of gas at a constant temperature is claimed to be described by Boyle's law, $PV = K$, where K is a constant—and also by van der Waals' equation, $(P + a/V^2)(V - b) = c$, where a, b, and c are constants. Obviously they both cannot be correct. However, when the sample of gas is of sufficiently low density, the term a/V^2 will be insignificant when compared with P, and V will be much larger than b. Then the product PV will be sufficiently similar to the result given by van der Waals' equation so that Boyle's law can be said to adequately describe the situation. However, at high densities van der Waals' equation must be used, but even it is not exact.

Occasionally and without warning you may be asked to investigate a law outside its limits of applicability so that the limits will be learned by personally discovering their existence.

There are several things to keep in mind when testing a predicted law or suspected law in a laboratory experiment. For example, in an experiment to see if a certain object falls with constant acceleration, the effects of air resistance might not be taken into account. The result would be expected to differ a little from that which was predicted, but the acceleration may vary only slightly as the object falls.

Further, the law itself may be only approximate in the form being studied. If this is the case, the results would not be exactly as predicted by the law, but would show how accurately the law describes the situation.

The results, even if all the conditions were perfect, could still not be expected to be exactly as predicted because no physical measurement is ever exact. There are always uncertainties in measured quantities. Because of these uncertainties, there will be an uncertainty in any result calculated from measured quantities. If this calculated quantity, considering its uncertainty, agrees with the results predicted by the law, the law can be said to be satisfactory *within the conditions of the experiment and within the range of error indicated by the uncertainty in the calculated answer.*

Even small discrepancies between the values predicted by a theory or law and the values actually measured can have enormous significance. For instance, the atomic masses of various elements once seemed to be integral numbers when compared with the standard for atomic masses which was then taken as one-sixteenth the mass of an atom of O^{16}. Some of these atomic masses are illustrated in the second column of Table 2-1, and it would be tempting to say from these data that the masses are integral numbers. However, in precise work, such as measuring individual isotopes as shown in column three of Table 2-1, the values obtained are those listed in column four.[8] These are based on $C^{12} = 12$ units. On the small differences between the actual masses and the integral numbers is based the principle for the release of energy by the nuclear processes of fission and fusion. It is a part of this difference between actual mass and integers that is converted to energy in nuclear reactions such as in hydrogen bombs, fission bombs, or nuclear reactors.

TABLE 2-1   A Few Atomic Weights and Nuclidic Masses Showing the Small Variation From Whole Numbers.

Element	Atomic Weight	Nuclide	Nuclidic Mass*
Hydrogen	1.01	H^1	1.00782522
Helium	4.00	He^4	4.00260361
Lithium	6.94	Li^7	7.0160053
Beryllium	9.01	Be^9	9.0121858
Carbon	12.01	C^{12}	12.0000000
Nitrogen	14.01	N^{14}	14.00307438
Oxygen	16.00	O^{16}	15.99491494

*These values were taken from Everling, et al.[8]

An important fact, too, is that the atomic masses are close to integral numbers, for this indicates that the nucleus is built of particles of roughly equal mass. From this generalization grew the concept of the nucleus being formed of protons and neutrons.

A general law involves an extension from experimental results. It may be found by experiment that something varies inversely as the 2.05 power of something else. It is then tempting and quite legitimate to say that the power is, in reality, probably 2, which is to say that an inverse square law describes the situation. The difference between 2.05 and 2.00 may be accounted for by experimental error or it may not. At any rate, it could be said that the basic law is probably inverse square and any real discrepancy from this would be accounted for by a secondary factor.

Laws are usually investigated or found in laboratories, although it is possible to do similar work outside laboratories. There are two main reasons for doing the work in the laboratory. One reason is that the factors influencing the situation under study will be more readily controlled in the laboratory, and another is that it is more convenient and often less expensive to work on the laboratory model. For example, two conditions under which waves can be studied are illustrated in Figure 2-1. Figure 2-1A shows waves on a lake on a stormy day. After the waves pass between the islands they are diffracted into the region behind the islands. Figure 2-1B shows waves in a small tank in a laboratory. These waves are diffracted into the regions behind the obstacles in the same manner as the waves going behind the islands. With the laboratory arrangement the wavelength can be changed quite readily, but in the natural situation one must wait for a day with a different wind speed to change the wavelength. In the laboratory setup the wavelength and velocity can be more easily measured than in the natural situation. The laboratory equipment is also less expensive (one ripple tank as contrasted to one aircraft), and the operating expenses may be considerably less in the laboratory.

In addition to the wave phenomenon already described there are also others. The grooved surface of a record acts as a reflecting type of diffraction grating, giving multiple spectra when light is reflected from it. Soap bubbles and oil films show colored areas and bands which are the result of the interference of light waves reflected from the two surfaces of the thin film.

Laws describe the order we find in nature, and when several laws concerning one particular thing have been found, a theory or model is devised to explain the laws. These theories or models often predict new phenomena that can then be investigated. Often there will be rival theories, each explaining the same set of laws. For example, by the time of Newton, laws concerning reflection and refraction of light had been found. These laws could be explained either on the basis that light was a stream of particles or that it was a wave. According to the wave theory, light should pass around corners as does sound or a water wave (Fig. 2-1). Since this was not observed by Newton, he was reluctant to accept a wave theory of light. With the demonstration of the diffraction of light around corners by T. Young in 1803, the wave theory of light became accepted and was not disputed until Planck, in 1900, introduced again the concept of a particle or photon. This concept, which amounts to a new particle theory, was shown by Einstein to be necessary to explain the photoelectric effect, or the interaction between light and electrons. Now we attribute both wave properties and particle properties to these photons. Note that through it all, while man changed his

A

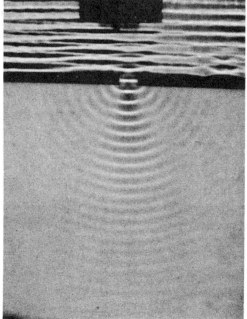

B

Figure 2-1 Two ways to study diffraction of water waves: **A,** Diffraction of waves passing between islands on a lake on a windy day. Note how the waves spread out in circles from the opening between the islands. (**A,** by the author.) **B,** The same phenomenon as in **A** but produced in a ripple tank in the laboratory. (**B,** Courtesy of Educational Services, Inc.)

explanation from particle to wave to dualistic theories, the empirical laws of reflection and refraction remained unaffected.

Concerning the rise and fall of theories, Poincaré remarked,

> Today the theories are born, tomorrow they are fashion, the day after tomorrow they are classic, the fourth day they are superannuated, and the fifth they are forgotten. But if we look more closely, we see that what thus succumb are the theories, properly so called, those which pretend to teach us what things are. But there is in them something which usually survives. If one of them has taught us a true relation, this relation is definitively acquired, and it will be found again under a new disguise in the other theories which will successively come to reign in place of the old.[9]

The theories "which pretend to teach us what things are" are the vulnerable things, though even theories are kept if they demonstrate certain phenomena to a sufficient degree of accuracy. The Bohr model or theory of the atom (a central positively charged nucleus with electrons revolving in certain allowed orbits) is useful in explaining the mechanism of the photoelectric effect or the general features of line spectra. In some calculations this model is satisfactory; in others it is not.

The criterion for the keeping of a law or a theory is its usefulness. When we have learned the limits, we can use the laws or theories with confidence inside their limits.

Another striking example of this process stretched out in time concerns the orbits of the planets. Tycho Brahe, the Danish astronomer (1546–1601), was a painstaking observer. His life was devoted almost entirely to measurement of the location of the stars and planets on the celestial sphere. Johann Kepler (1571–1630) came from Germany to work with Tycho Brahe just a year before Brahe died. Kepler continued the gathering of data and from them was able to formulate his three laws of planetary motion. These dealt with the ellipticity of the orbits, the law of equal areas, and the relation between the period of a planet and the size of its orbit. These laws had a completely experimental basis. Sir Isaac Newton (1642–1727) made use of Kepler's laws of planetary motion in the formulation of the theory of a universal gravitational force between all masses and then in the formulation of what we know as the law of gravity. Newton's concept of a force acting at a distance was a bold and useful hypothesis. According to his gravitational theory, even masses in the laboratory attracted each other, and such an attraction was detected and measured by Henry Cavendish in 1798. We may consider gravity as a real force, but it is not necessary to do so in order to explain the observed motion of objects. Consider that you are looking down into a darkened room at a table, in the center of which is a large ball. You see a person enter and place a small ball on the table. The small ball accelerates toward the central one. Then the person pushes a small ball sideways, and it orbits the central one. These motions can be explained by saying that there is an attractive force between them. This is just how Newton described or explained falling objects and planetary motion. The force he called gravity.

But there is another explanation. Perhaps the table is flexible and the central mass distorted the surface into a shape somewhat like that of a bowl. A small mass put near the center rolls into the central mass; a small mass given a sideways push orbits it. (You should try this with a mixing bowl.) The motion observed from above can be said to be the result of the distortion of the table by the presence of the mass. This is in effect what Einstein did in the general theory of relativity. He dispensed with the force acting at a distance and substituted curved space. Einstein's concept explained all the phenomena or laws of motion that Newton's did, but it also explained some phenomena that Newton's did not, the motion of the orbit of the planet Mercury being one.

Except for the deviance of Mercury, all the laws of planetary motion were unchanged when the concept of what was behind the laws changed from action at a distance to the warping of space by matter. It is the theories or models that change, not the laws. Paul Heyl put it very nicely when he said,

> No consequences, however transcendental, can alter the fact that the theory of relativity is but a working hypothesis, constructed to fit more closely the experimental facts of Nature than Newton's hypothesis proved able to do.[10]

Because in most instances Newton's gravitational theory gives sufficiently precise results, and because it is so much simpler to work with than is the general theory of relativity, the Newtonian concepts are still used. Satellite orbits, first considered by Newton, are still calculated on the basis of his description of gravity.

In this example there are all steps in the process: the gathering of data by Brahe and Kepler, the formulation of laws by Kepler, and the proposal of a theory by Newton. Newton's theories of gravity were useful in many ways, including the prediction of the force which Cavendish measured. The concept of a gravitational force is still useful even though it has basically been superseded by general relativity theory.

Problems

1. According to the theory of relativity the mass of an object in motion with a speed v is given by

$$m' = \frac{m_0}{\sqrt{1 - \frac{v^2}{c^2}}}$$

where m_0 is the mass when it is at rest and c is the speed of light. Newtonian mechanics does not consider any such variation in mass with velocity. Calculate the mass of an object which is moving at a speed of 8 km./sec. (approximate orbital velocity for a low earth satellite) if its rest mass is 1 kg. At satellite speeds would Newtonian mechanics be expected to be satisfactory? A binomial expansion could be used in the calculation of the increase in mass.

2. Compare the constants given by Boyle's law with those given by van der Waals' equation for 1 liter of oxygen at 1 atm. of pressure. The product

PV (Boyle's law) is 1 liter atm. According to van der Waals' equations, $(P + a/V^2)$ $(V - b)$ is constant. For oxygen, a is 0.00271 and b is 0.00142 in the appropriate units. Evaluate the constant from van der Waals' equation and find by what percentage it differs from the product *PV*.

3. Make a table of your data from the experiment on the period and amplitude of a simple pendulum, arranging the measurements in order from small to large amplitude. Then make a graph of the data. You may notice a trend in the tabulated measurements, and this will be made more clear by the graph. Discuss the results with respect to the idea of physical laws being valid only within a specified limit.

EXPERIMENT

2-1

Apparent Depth

It is a familiar phenomenon that when one looks at something in water it does not seem as far down as it really is. This apparent decrease in depth is particularly noticeable in swimming pools, and even in glasses or cups of water. As well as having uses in scientific measurements, it is also used in some parlor tricks. You might wonder how much closer the bottom of a cup appears when water is poured into it. In other words, how is the real depth related to the apparent depth?

The phenomenon is important in the use of a microscope as a measuring device. The thickness of an object is sometimes measured by focusing on the bottom of an object and then on the top. The distance that the microscope moves is a measure of the thickness or vertical distance, but because of the apparent depth phenomenon, the true distance is not given directly. Measurement of the length of tracks of nuclear particles in photographic emulsions or in bubble chambers is an example of the application of this method in physics. The phenomenon is also important in microscope work in biology and even in underwater photography. It should be noted, however, that with oil immersion microscopy this phenomenon becomes less important but more complex.

The project should be done in four parts:

A. Find the ratio of real depth to apparent depth for water.

B. Find the ratio of real thickness to apparent thickness of a glass block in air.

C. Find the ratio of real thickness to apparent thickness of a glass block covered with water.

D. Analyze the apparent depth phenomenon theoretically.

A. A measuring microscope is set up with the axis of the microscope parallel to the scale and is arranged to look down into a beaker in which water will be put. The apparatus is illustrated in Figure 2-2. A mark on which to focus the microscope must be put on the bottom of the beaker. One way to do this is to stick a small piece of masking tape to the bottom of a beaker on the inside and make some pencil marks on the tape so that the microscope can be accurately focused on it. Place the microscope above the beaker. (The microscopes have a working distance of about 4 inches if a two-power objective is used.) Focus on the tape and determine the microscope scale reading r_1. *Remember to keep the focusing knob (if there is one) fixed* and to move the microscope tube up and down by means of the knob that moves the microscope along the scale. Pour some liquid into the beaker so that about 1 cm. of liquid covers the bottom. Focus on the tape again and determine the microscope reading r_2. Put some chalk dust or bits of paper upon the surface of the liquid and when these are in focus, determine the microscope reading r_3. Record all readings directly into a table. The real depth, d, of the liquid is just the difference between r_3 and r_1. The apparent depth, d', is the difference between r_3 and r_2. By increasing the liquid level by about 1 cm. each time, obtain five or six sets of readings for r_2 and r_3. Calculate d and d' and the ratio of the real depth to the apparent depth in each case.

B. Repeat part A, but with a glass block rather than with a liquid. To find the real thickness a caliper may be used, and the apparent thickness is found as with the water. Calculate the ratio of the real thickness to the apparent thickness of the glass.

C. Repeat part B, but with the glass block in a beaker and covered to any depth with water. Do not, however, let the microscope go into the water. Compare the ratio of real to apparent thickness in this case with that found in part B.

D. Parallel to the experimental investigation is the theoretical analysis. The bending of a light ray at a surface is described by Snell's law, $\sin \alpha / \sin \beta = \mu$, in which the angles α and β are as shown in Figure 2–3. The index of refraction is μ. The tangents of α and β can be written in terms of x and the apparent and real depths. For small angles such as are encountered in viewing normal to the liquid surface, the tangents and the sines are almost identical. Carry this analysis through and solve for the ratio d/d'. Use tables to see if your experimental result and the results of the theoretical analysis agree.

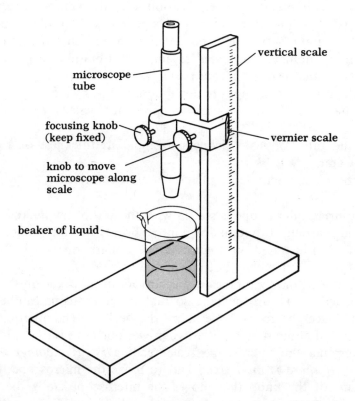

Figure 2–2 The apparatus used to measure the apparent depth of a solution.

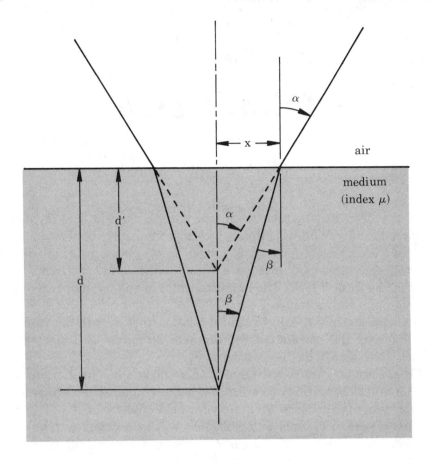

Figure 2–3 A diagram to illustrate the theoretical analysis used to determine the ratio of real depth to apparent depth. The angle α will be assumed to be very small.

In this project you will have found the law relating real depth to apparent depth, or real thickness to apparent thickness. Initially you found a constant relation, and then, either through your knowledge of optical constants or after the theoretical analysis, the meaning of the constant was found.

Apparatus

Measuring microscope with a 2 × objective lens
Beaker — 100 to 500 ml.
Piece of glass at least ¼ inch thick
Micrometer caliper
Water or other transparent liquid
Tables of physical constants
Vernier caliper or micrometer.

2-2

Ohm's Law

Ohm's law describes the relation among voltage, current, and resistance in an electric circuit, and is without a doubt the most frequently used of all the laws in the field of current electricity. If a voltage V is applied to a circuit having a resistance R, the current I that flows is found with the use of Ohm's law, often expressed in the form $V = IR$. The units are volts for voltage, amperes or amps for current, and ohms for resistance.

If you make a graph of voltage against current, that is, with the voltage on the abscissa or y axis and current on the ordinate or x axis, a straight line would result. The slope would be the resistance R.

In this project you are to use Ohm's law to make some predictions about the current that will flow through a resistor of a given value, after which you are to measure the current to see how accurate your predictions were.

The first object to be used as a resistance will be a common type of resistor used in electronic circuits, either a carbon type or a wire wound type. The value in ohms will be marked on the resistor either as a number or as a series of colored bands. If there are colored bands, the first band indicates the first digit, the second band the second digit, and the third band the power of ten by which to multiply to get the resistance in ohms. The color code is based on the following:

black = 0	green = 5
brown = 1	blue = 6
red = 2	violet = 7
orange = 3	gray = 8
yellow = 4	white = 9

Having found the resistance, make a table with columns for voltage (from 0 to 10 in steps of two volts), the calculated current [you may need to use milliamps, where one milliamp (ma) is 0.001 amp], and the measured current.

The circuit to use is illustrated in Figure 2-4. First connect the voltage source, rheostat and voltmeter, and obtain a variable voltage by moving the tap on the rheostat. Then connect the milliammeter and the resistance. Make the required readings, and determine by what percentage they differ, on the average, from the calculated values.

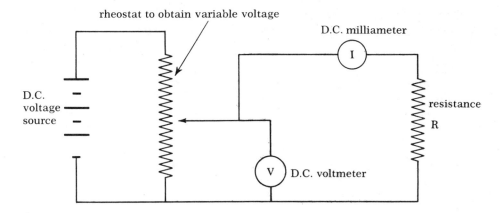

Figure 2-4 The circuit diagram for the study of Ohm's law. In Experiment 2-2 the item marked R will be first a carbon resistor, then a water cell, and finally a germanium diode.

Also make a graph of your readings to see how accurately the project has shown Ohm's law to apply.

The next step is to study the application of Ohm's law to other objects. In place of the resistance in the circuit of Figure 2-4, put a beaker of slightly acidified water with two wires or plates in it. Measure and tabulate V and I. Plot them on a graph. Does Ohm's law seem to apply?

Before worrying too much about this, replace the beaker of water with a germanium diode. First connect it in one direction and make a series of readings. Then connect it in the other direction. Graph these data also.

You may have found that Ohm's law in its simple form is restricted in its use; but where it does apply, it applies quite accurately. The discrepancies found could be accounted for by saying that the resistance changes and the actual value can be defined by

$$R = V/I$$

Alternatively, the resistance can be defined as the slope of the graph V against I. This is often called a differential resistance.

Using either of these definitions (but state which one), comment on the behavior of the resistance of a beaker of water. Also calculate an approximate "forward" and "backward" resistance for the diode.

What about other objects? Replace the resistance with your lab partner, who should hold the wires with moistened hands. You may have to replace the milliammeter with a microammeter. Do not raise the voltage too high. What resistance does your lab partner offer?

Try using a small light bulb, such as one from a Christmas tree or a flash light; but with the latter, take care not to exceed the rated voltage.

Apparatus

Direct voltage source up to about 10 volts
Rheostat — about 100 ohms
Resistor — about 100 ohms

Multimeter (or milliammeter and microammeter)
A second multimeter or a voltmeter — 0–6 volts D.C. (at least 10,000 ohms per volt)
Beaker
Metal plates to fit in beaker
Crystal diode

2-3

The Bouncing Ball

It can be shown that when two perfectly elastic objects collide, the speed of separation will be the same as the speed of approach. If the objects have no tendency at all to return to their original shape after they are deformed in a collision, they do not separate at all. Such a collision is described as being *perfectly inelastic.* The measure of elasticity in a collision is the ratio of velocity of separation to velocity of approach, and this ratio is called the *coefficient of restitution,* designated by e:

$$e = \frac{\text{velocity of separation}}{\text{velocity of approach}}$$

If $e = 1$, the collision is perfectly elastic.

If $0 < e < 1$, the collision is partially elastic.

If $e = 0$, the collision is perfectly inelastic.

The purpose of the first part of this project is to see if in a real situation the coefficient of restitution is indeed a constant, or, if it is not, to determine what amount of variation may be expected.

If a ball is dropped onto a hard surface from a height h_1, it hits at a speed v_1 which is equal to $\sqrt{2gh_1}$. The ball leaves the surface at a speed v_2, which makes it rise to a height h_2, and v_2 is $\sqrt{2gh_2}$. The velocity of approach is v_1, and v_2 is the velocity of separation. From these quantities, e is found. So the idea is to drop a ball from a measured height and find how high it rises on the first bounce. Repeat this procedure at a wide range of heights, and obtain the coefficient of restitution for each bounce. Organize the results, tabulate them, and comment on the law that states that in a collision between the objects, the ratio of the velocity of separation to the velocity of approach is a constant.

To simplify the calculation of all the velocities, use $v = 44.3 \sqrt{h}$, which will give v in cm./sec. if h is in cm.

A steel ball bearing dropped alongside a meter stick onto a smooth steel surface is a good combination to use. Also try a rubber ball or a plastic ball, the kind with supposedly superior bouncing characteristics. What is a measure of this?

From what part of the ball should you make the measurements?

In the conclusion for this part of the project, summarize your data and write a paragraph or two about the limits of that law about the perfection of the elasticity of whatever materials you used.

The second part of the project is an optional extension, and it is largely theoretical. Can you answer the questions:

1. How many bounces will the ball make before it comes to rest?

2. How long will it take the ball to come to rest after being released from a height h_1? To answer this, calculate the time to fall h_1 (call it t_1), then the time to rise to h_2 and fall again (t_2). It rises to a height h_3 and falls back in a total time t_3, etc. This can be approached theoretically using your mean value of e to calculate the height of each successive bounce. The times t_1, t_2, t_3, and so forth are added; you will see that a series results and that the series has a limit. The ball may bounce an infinite number of times, but in a finite amount of time. If you have been able to calculate this, see also if you can obtain an experimental measurement of it.

Apparatus

Steel ball bearings
Heavy smooth steel plate
Meter stick
Rubber ball
Plastic ball
Stop watch (optional)

EXPERIMENT

2-4

Lenses and Images

A law has been described in the beginning of this chapter as the description of a relation between two quantities that occur in a phenomenon, often expressed as a mathematical formula. You should read the first part of the chapter before proceeding.

This project involves the formation of an image with a lens and the discovery of a relation between object and image positions, although in a rather uncommon manner. Instead of being made from the lens position, the measurements will be made from the focal points or, more properly, the "principal foci." The reason for doing this is that the lens formulae usually presented are for thin lenses, ones for which the thickness is negligible. In practical situations lenses used are rarely thin, and the point from which object and image distances should be measured is not clear. This problem disappears when the measurements are made from the focal points, and the resuiting relation that you will find is applicable also to thick lenses. The relation to be found is not new; it was discovered by Sir Isaac Newton, but has since been much neglected. It is rarely even mentioned in texts, although it can be very useful in practice.

The suggested procedure is to mount a lens on an optical bench, and, by a method described later, to mount pointers at the principal focal points. Then you mount an object (a lamp or an illuminated scale or picture) on the optical bench on one side of the lens and a screen at the position of the sharpest image on the other side. Then measure the distances a and b as shown in Figure 2-5. Obtain a variety of measurements of values of a and b, and try to find how a and b are related. The statement of this relation could be called a law.

For example, Kepler measured mean orbital sizes R for the planets and the time T for each to circle the sun. He discovered that for each of them the ratio R^3/T^2 was the same, or constant. That $R^3/T^2 = K$, where K is a constant, is called Kepler's third law. Robert Boyle found that with an enclosed mass of gas the product of measured values of pressure P and volume V was a constant; the relation $PV = K$, where K is a constant, is called Boyle's law. You are to find a combination of a and b that yields a constant value, and the mathematical statement of this is effectively a law.

Some hints in establishing the principal focal points may be of value. A principal focus is the place at which light parallel to the axis of the lens is focused. Conversely, light diverging from that point is focused into a parallel beam. If an illuminated object such as a pointer is placed at the focal point, light diverging from it is focused by the lens to a parallel beam. If a plane mirror is placed to reflect the light back through the lens, it comes back as a parallel beam and is focused to the focal point. The object and image are in the same position! To

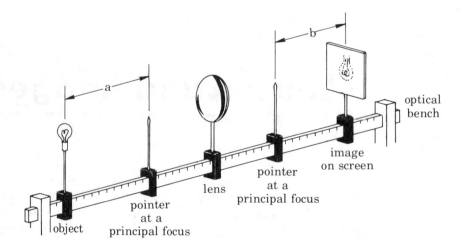

Figure 2–5 The arrangement of the apparatus to find the relation between the object distance and image distance, each being measured from a principal focal point.

locate a focal point, mount a pointer on one side of the lens and a plane mirror on the other, as in Figure 2–6. Look past the pointer toward the lens to see the image, and move the pointer back and forth until its image is inverted and the same size as the real pointer. Adjust the position of the pointer slowly until it coincides in space with its image. This has occurred when, if the pointer and its image are seen point-to-point, as your head is moved back and forth they remain point-to-point, or do not move relative to each other. We say that there is no parallax between them. To understand this a little better, place a small light bulb on one side of the pointer and a white paper on the other side. The image will form on the paper.

Leave the pointer at that position, and repeat the procedure with another pointer to determine the focal point on the other side.

You are now ready to complete the work as suggested.

When you have found the equation asked for, make an approximate measurement of the focal length f – the distance from the lens to a principal focus – and determine how the constant is related to f. The constant is, in fact, a simple

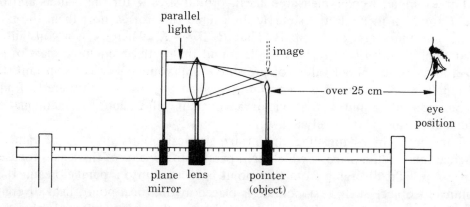

Figure 2–6 The arrangement of apparatus to position a pointer at a principal focal point. When the pointer and its image are at the same distance from the lens, the light between the lens and the mirror is in a parallel beam.

power of f. Because the lens is thick, f cannot be measured precisely, but the power involved should be obvious.

Your result will be presented as the equation showing a combination of a and b that yields a constant and a statement of how that constant is related to the focal length.

You may be able to derive the relation using a ray diagram and perhaps with the assistance of your text, but be sure to do the experimental work first. If you do this, does your experiment validate your theoretical analysis?

An extension of this project involves the use of a camera as a measuring device. The film is ordinarily at the principal focus, and to photograph a nearby object the lens is moved forward by an amount b, which is an easily measured distance. The focal length is invariably marked on a camera lens. A few more measurements with your lens and optical bench will show that the ratio of image size to object size is the same as the ratio b/f. An object size can be calculated using the measured image size on the film and a measurement of b at the time the photograph is made. This will allow measurements in the range of 2 to 5 per cent accuracy.

Apparatus

Optical bench with mounts for lens, two pointers, plane mirror, object and screen

Lens of focal length 5 to 15 cm.

Object — small light bulb or illuminated scale or picture

Screen — white card or ground glass

Plane mirror

Meter scale

A camera and film for the optional part

REFERENCES

1. Poincaré, H.: The Value of Science. New York, Dover Publications, 1958, p. 85.
   Poincaré has been called one of the last of the universalists. He was a superb physicist, mathematician and philosopher of science. It is only rarely that a working scientist expounds on the philosophy of science, and Poincaré did this at a time when classical physics was showing its inadequacies and Einstein was bringing out the new concepts of relativity.
2. *Ibid.*, p. 7.
3. Newton, I.: Opticks Ed. 4. New York, Dover Publications, 1952, p. 404.
   Sir Isaac Newton wrote two books, the *Principia Mathematica* (in Latin, but now available in an English translation) and *Opticks.* This now readily available book contains many fascinating statements and ideas expressed lucidly, as could be done only by one of the greatest minds of all time.
4. Wilkinson, D. H.: *In* Turning Points in Physics. Amsterdam, North Holland Publishing Company, 1959, p. 1964.
   This book is a collection of essays, initially presented as a series of lectures at Oxford University. The authors are all top men in the field about which they write and also are remarkably proficient at expressing their ideas. The material is from 1958, but most of it is still valid.
5. Poincaré, p. 130.
6. *Ibid.*, p. 77.
7. *Ibid.*, p. 3.
8. Everling, F., König, L. A., Mattauch, J. H. E., and Wapstra, A. H.: Relative nuclidic masses. Nucl. Phys. *18*:529 (1960).

9. Poincaré, p. 139.
10. Heyl, Paul R.: Rapport, S., and Wright, H., (eds.): Physics. New York, New York University Library of Science Series, 1964, p. 313.
   This book is a collection of essays by many of the great physicists who made major contributions to the development of physics. The contributors include Einstein, Gamow, and Bragg, and their own words are well worth reading.

3

AN EXACT SCIENCE

Exactly! What does this word mean? Has anyone ever made an exact measurement? The word *exactly* does, of course, have a meaning (look it up in a good dictionary), and it has its uses in science. The yard is defined in the following way: ". . . the international yard equals 0.9144 meters . . . making one inch exactly equal to 2.54 centimeters."[1] This is a definition and the word *exactly* can be used, but when applied to measured quantities or to values calculated from measured quantities the word *exactly* cannot be used. The reason is that there is no such thing as an exact measurement. Cannot a simple quantity like the length of this page be determined exactly? To measure it, one could use a scale (rule). The scale would be laid along the page and observations made at the two edges of the page. Try this. One edge may be set at zero, but can you set it exactly at zero? Would a magnifier help you to set it more precisely? A microscope would be even better, but the zero position of the ruler may be marked by a line, and the line is probably too wide to set accurately by means of a microscope. At the other edge of the page similar difficulties will be encountered. Even if one used a magnifier or some other device, the exact reading of the scale could not be determined.

Another difficulty is also encountered. How accurate is the scale being used? One wooden scale may not indicate the same distance as another, similar wooden or steel scale. Each scale is the end result in a chain of copies of the international standard meter. Furthermore, the length of a scale may vary with such factors as time, temperature, or humidity. No scale itself is exact and a degree of precision may be theoretically achieved at which no two scales would give the same reading of the same object.

As if this were not enough, there is a further difficulty. Is the length of the page the same at all positions? No physical object can be made perfect, so there are always variations inherent in any object. Often the variations will be undetectable by the measuring instrument being used. If we used a vernier caliper, the variations in the diameter of a metal rod might be undetectable, but a more precise instrument would show that the diameter was not exactly the same at all places. In other instances the variations would be readily apparent. A relatively soft ball of a material such as lead which had been subjected to years of use in a physics laboratory would undoubtedly show variations in its diameter when measured in different directions with a vernier caliper. The diameter of the ball would not be a meaningless quantity, however, for a large number of measurements could be made and the average diameter could be determined. The average diameter could not be stated exactly because even one more measurement would change the average. However, the larger the number of measurements, the more precisely the average diameter would be determined.

Exact measurements, then, are not possible. Each measurement is, however, determined to a certain precision. The precision of the measurement depends on: (a) the instruments and method used in making the measurements; (b) the precision of standardization of the measuring device; (c) the actual variations in the measured quantity; and related to this, (d) the number of determinations made.

When we ask the question, How inaccurate is a certain measurement? or, How precisely do we know a certain quantity? we are referring to what is called the *error* in the quantity. By the word *error* we do not mean "mistake" but rather the uncertainty in a quantity caused by all the contributing factors. Every measurement has an error. Every calculated quantity which is based on measured values has an error.

The error may be expressed in either of two basic ways, which are referred to as the *absolute error* and the *relative error*. The absolute error is expressed in the units of the measured quantity. The relative error is expressed either as a fraction of the measured quantity or as a percentage of the measured quantity. It is customary to indicate the uncertainty, or the error, in a quantity by a ± after the quantity, viz.:

$$23.1 \text{ cm. } \pm 1\%$$

or

$$23.1 \pm 0.2 \text{ cm.}$$

These are respectively percentage or relative error and absolute error.

COMPARING TWO QUANTITIES

The error in the calculated result of an experiment can be as important as the answer itself. Experiments will often involve determining a quantity by two different methods and then seeing if the results agree. Sometimes a theoretically expected answer will be compared with the corresponding experimentally measured value. It will be rarely, and then only by chance, that there will be exact numerical agreement, so the errors in the two quantities being compared must be obtained. Then, if the ranges of the two values overlap, they may be said to agree within the limit of error of that experiment. If they do not overlap, then the two values do not agree.

We cannot say that 19.9 is the same as 20.5, but if these two numbers each have an error such as 19.9 ± 0.3 and 20.5 ± 0.4, then the first value is in the range 19.6 to 20.2 and the second value is in the range 20.1 to 20.9. These ranges overlap, so it may be said that these numbers agree *within their errors.* Another way of describing the comparison is to say that the two values have *not been shown to be different*. If the numbers with their errors had been 19.9 ± 0.2 and 20.5 ± 0.2, the two would differ by more than their experimental errors.

On September 12, 1959, the rocket Lunik II was launched on a path destined to impact it on the moon. The huge radio telescope at Jodrell Bank in England was trained on the rocket, and the radio signals told indirectly of its increasing speed as it fell to the moon. From the speed, the distance from the center of the moon could be computed. Suddenly, on September 13th at 21 hours 02 seconds universal time, the signals ceased. Computation from the radio data showed the rocket to be still 70 km. above the surface of the moon. Newspaper reports at the time used this result as the basis for a claim that the whole thing was a hoax and that the signals were false. But the moon is about 385,000 km. (240,000 miles) away. How accurately could the position of the rocket be calculated when it was at that distance? The error was worked out[2] to be ± 150 km. The distance from the surface of the moon at the time of cessation of the signals was, therefore, not 70 km. but more properly 70 ± 150 km. In other words, the rocket was determined to be somewhere in the region 220 km. above to 80 km. below the surface of the moon, and that is the best that can be said. This is not evidence that the rocket did not reach the moon; the inclusion of the error with the measurement could have prevented useless arguments.

"ROUNDING OFF" ERRORS

In any measured quantity there is an error as a result of rounding off the number of the last figure quoted. If a length is quoted as 10.3 cm., the implication is that the actual length lies between 10.25 and 10.35 cm. The error is ± 0.05 cm. If the value had been 10.30 cm. (that is, between 10.295 and 10.305) it would have been necessary to record the last zero to indicate the precision of the measurement. If a number is written as 186,000, which is the velocity of light in miles per second, there is no indication whether the zeros are space fillers. A more precise value would be 186,283 miles/sec. If the number had been expressed in powers of 10—that is, in the form 1.86×10^5 miles/sec. — there would be no confusion about the amount of rounding off. The corresponding value in the metric system is 300,000,000 meters/sec. This should be written as 3.00×10^8 meters/sec., for the first two zeros are significant. The numbers 1.86×10^5 and 3.00×10^8 are said to be accurate to three significant figures. In four significant figures the velocity of light is 1.863×10^5 miles/sec. or 2.998×10^8 meters/sec.

In general, most measurements in introductory physics experiments are to three significant figures. Occasionally, four figures will be obtained and sometimes only one or two figures.

Incidentally, the number represented by π is, to eleven significant figures, 3.1415926536. To four figures it is 3.142 and to three figures, 3.14. The quantity 22/7 is sometimes used for π; and this, expressed as a decimal, is 3.142857. This is obviously not equal to π and even rounded off to four figures, 22/7 is 3.143, which is significantly different. When rounded off to three figures, however, it does agree. The difference amounts to 0.4 per cent and this error is of such a size that it is advisable to develop the habit of using π in a decimal form rather than as 22/7. A number like π should be used in a form which is of higher accuracy than the measurements so that no significant error is introduced into the answer because of the rounding off of such a number. Tsu Ch'ung-chih, in China about the year A.D. 260 on our calendar, found that the fraction 355/113 was a closer approximation of π than was 22/7 and, in fact, it differs from π by less than one part in ten million. Very rarely will a calculation require a more precise value than this, but still it should be remembered that it is an approximation.

ERRORS IN DEVICES TO MEASURE LENGTH

The Wooden Scale. The student is undoubtedly familiar with the use of an ordinary ruler to measure objects, but a few details should be mentioned. The length of a wooden scale may vary considerably with time, and it is not uncommon to find a meter stick that has changed in length by 2 mm. or 0.2 per cent. Unless the meter stick has been calibrated against a laboratory standard, an error of ± 0.2 per cent should be considered in any quantity measured with a wooden scale because of the uncertainty in the accuracy of the scale. If the scale has not been made by a reputable firm, the error may be much more than this.

If the zero end of the rule looks at all battered or inaccurate, the scale should be placed across the object being measured and readings should be obtained at the two ends of the object. The distance can then be found by subtraction.

Errors resulting from parallax in reading a scale can be reduced by viewing past the edge of the scale so that the line of sight is perpendicular to the scale or, if convenient, by placing the meter stick on edge so that the lines on it make contact with the object.

The total error in a quantity measured with a wooden rule is the calibration error of 0.2 per cent and the error in reading, which is often ± 0.5 mm. Of course, to add the two errors they must be expressed as absolute errors.

The Steel Scale. A steel scale made by a manufacturer of scientific apparatus will usually be correct to within a tenth of a millimeter in 1 meter. In other words, it will have an error of about 0.01 per cent, which is usually negligible compared with errors in the reading of the scale. With care in reading the scale, measurements with a steel scale may be obtained to about ± 0.2 mm. depending upon the experience of the observer and considering an error at both ends of the object being measured.

The Vernier Caliper. The principle of the varnier caliper is described in Chapter Ten. To measure an object with a vernier caliper, the object is placed between the jaws and a reading is taken. Then the object is removed and the jaws are closed to obtain a zero reading. Very frequently a vernier caliper will register one or two tenths of a millimeter on one side or the other of zero when it is closed, and this zero correction must be properly applied. When a measurement is made, the observed reading, the zero reading, and the final measurement should all be recorded in the notebook. The vernier caliper commonly used in physics laboratories can be read to the nearest tenth of a millimeter, and considering the uncertainty in reading the measurement and the setting, the error in a measured quantity should not exceed ± 0.1 mm. A vernier caliper from a reputable scientific supply house should be graduated with sufficient accuracy that the error due to graduation is less than this amount and need not be considered.

The Micrometer Caliper. The micrometer caliper is described in detail in Chapter Ten. In using the micrometer, one must take care to have the same light pressure exerted for each measurement, because forcing the screw can easily bend the jaws open a small amount or compress the object being measured. Some micrometers are provided with a ratchet to assure that equal pressure is always obtained. If your micrometer caliper has a ratchet, always close it by means of the ratchet knob. A zero correction must always be made just as is done for a vernier caliper. The divisions on the thimble of the micrometer caliper often used in physics laboratories indicate 0.01 mm. The error in each reading is about ± 0.005 mm., and considering the error in the zero setting, an object may be measured to ± 0.01 mm.

ERRORS IN THE MEASUREMENT OF MASS

The error in the determination of a mass using a balance will originate in two places: first, in the calibration of the comparison weights and second, in the ability to determine the balance point. The latter error may result from difficulty in accurately reading the scale, from the inherent sensitivity of the balance, or from its sluggishness. Excessive sluggishness is caused by damaged knife edges, and with a sluggish balance small additions of weight may make no significant change in the balance position.

To get an idea of the precision of a balance, several weights known to be accurate should be weighed on the balance. Furthermore, if the balance is to be used with a variety of uncalibrated weights, a selection of these weights should be measured on the balance to determine the variation of the weights themselves.

ERRORS IN CALCULATED QUANTITIES

Measured quantities are not exact, so quantities which are calculated using measured values cannot be exact either. In a general way the calculated value can be no more precise than the measured numbers; in fact, the errors accumulate so the calculated value is usually less precise than the measurements on which it is based. Getting an answer correct within 1 per cent requires measurement of a precision greater than 1 per cent.

The error in a calculated quantity can be determined from the errors in each of the quantities used in the calculation. This will be illustrated first in a very approximate way, using a method based on significant figures, and then more precisely, using a method based on the numerical value of the error in each quantity used in the calculation. The treatment will not have the rigor of statistical analysis, but it will serve to give an estimate of the precision of a result. Each of the mathematical operations—addition, subtraction, multiplication, and division—will be considered.

Consider the case in which two objects are to be placed together on a scale pan. Each has been previously weighed and the weights were 19.3 gm. and 1.52 gm. What would the combined weight be? The 19.3 gm. object is not exactly 19.3 gm. but the weight is between 19.25 gm. and 19.35 gm. The first unrecorded figure may have been 0, 1, 2, 3, 4, 5 or −1, −2, −3, −4, −5. The first unrecorded figure is an unknown number which could be represented by a question mark; and, in fact, each following position could be represented by a question mark. More precise measurement would determine the numbers to replace the question marks. The sum being obtained is then:

$$
\begin{array}{r}
19.3?? \text{ gm.} \\
+ \quad 1.52? \text{ gm.} \\
\hline
20.8?? \text{ gm.}
\end{array}
$$

The sum is 20.8 gm., with all numbers after the 8 being unknown because the sum of 2 and an unknown number is unknown.

In some instances one number is too small to add to another. In a microscope a rod may be seen to expand by 0.025 cm. when it is heated. If the original length was measured and found to be 75.2 cm., the length after heating is:

$$
\begin{array}{r}
75.2??? \text{ cm.} \\
+ \quad 0.025? \text{ cm.} \\
\hline
75.2??? \text{ cm.}
\end{array}
$$

The expansion makes no significant difference in the length, even though the expansion was measured.

The following examples will show that a similar reasoning applies for subtraction:

$$
\begin{array}{r}
31.4?? \\
- \quad 3.63? \\
\hline
27.8??
\end{array}
\qquad\qquad
\begin{array}{r}
75.2?? \\
- \quad 0.025 \\
\hline
75.2??
\end{array}
$$

It is advisable to round off to the same decimal position numbers which are to be added or subtracted; for example, 33.6 minus 2.67 should be changed to 33.6 minus 2.7, giving the answer 30.9.

The technique of using question marks in place of the unrecorded digits can be used in multiplying and dividing also. Note that in the following examples each question mark is handled just like a number:

(a)
$$
\begin{array}{r}
4.12? \\
\times \ 2.31? \\
\hline
???? \\
412? \\
1236? \\
824? \\
\hline
9.51????
\end{array}
$$

(b)
$$
\begin{array}{r}
4.1234? \\
\times \quad 2.3? \\
\hline
?????? \\
123702? \\
82468? \\
\hline
9.4??????
\end{array}
$$

(c)
$$
\begin{array}{r}
1.34? \\
174?\overline{)233????} \\
174? \\
\hline
59?? \\
522? \\
\hline
7??? \\
696? \\
\hline
1????
\end{array}
$$

(d)
$$
\begin{array}{r}
2.7? \\
16?\overline{)44.33} \\
32? \\
\hline
12?? \\
112? \\
\hline
1??
\end{array}
$$

Example (a) shows that the product of two numbers, each having three significant figures, has only three significant figures. It is more correct, if the numbers are measured quantities, to say that the product of 4.12 and 2.31 is 9.51 than it is to say that the product is 9.5172. The latter number implies much higher accuracy than was actually achieved. In the example marked (b), the product of a five figure number multiplied by a number with only two figures has only two significant figures. Similarly, in (c) dividing a number with three figures by another with three figures gives an answer with only three figures. The general rule is that in multiplying or dividing, the answer may have no more significant figures than did the number with the least significant figures used in computing the answer. This is reasonable, for an answer can be no better than the worst measurement used in its calculation. A common analogy is that a chain is only as strong as its weakest link.

It is common to make readings to three significant figures because this accuracy is obtainable with ordinary measuring scales and balances. Answers then have only about three significant figures. It is for this reason that a slide rule is commonly used by physicists to compute results. There is nothing to be gained by computing answers in longhand or with four or five place logarithms if the measurements are accurate to only three figures.

THE CALCULATION OF ERRORS

ADDITION

Consider the sum of two numbers such as 75.3 ± 0.2 and 7.6 ± 0.4. The sum of the two numbers without the errors is 82.9, but how accurate is the 82.9? The number 75.3 ± 0.2 could be as small as 75.1 or as large as 75.5. The number 7.6 ± 0.4 lies between 7.2 and 8.0. The sum could be as small as $75.1 + 7.2$ or 82.3, and it could be as large as $75.5 + 8.0$ or 83.5. This range, 82.3 to 83.5, can be represented by writing 82.9 ± 0.6. The error 0.6 is the sum of the absolute errors in the two numbers being added.

In symbols we can let the number be represented by a and b and their respective absolute errors by Δa and Δb. The sum is represented by $(a \pm \Delta a) + (b \pm \Delta b)$. The sum could be as small as $(a - \Delta a) + (b - \Delta b)$ or $(a + b) - (\Delta a + \Delta b)$, and it could be as large as $(a + \Delta a) + (b + \Delta b)$ or $(a + b) + (\Delta a + \Delta b)$. This possible variation, from $(a + b) - (\Delta a + \Delta b)$ to $(a + b) + (\Delta a + \Delta b)$, can be represented by writing $(a + b) \pm (\Delta a + \Delta b)$. The quantity $(\Delta a + \Delta b)$ is the error in the sum $(a + b)$. This error, determined by adding the absolute errors in the quantities, gives the maximum error in the sum, and the error calculated in this manner is called the *possible error*.

The rule, then, is that when numbers are added, the absolute errors add.

SUBTRACTION

Errors add when quantities are added. What happens to the errors when quantities are subtracted? Consider the same numbers as were considered before, but this time subtract them; that is, 75.3 ± 0.2 minus 7.6 ± 0.4. The outer limits of the result would be found by taking 75.5 minus 7.2 and 75.1 minus 8.0. These outer limits are 68.3 and 67.1. Subtracting the numbers themselves gives 67.7, and the limits are 0.6 above this and 0.6 below this. The answer is then 67.7 ± 0.6. In symbols,

$$(a \pm \Delta a) - (b \pm \Delta b) = (a - b) \pm (\Delta a + \Delta b)$$

Thus, *when quantities are subtracted their absolute errors add.*

MULTIPLICATION

Multiplication can be used to find an area, and the uncertainty in an area depends on the uncertainty in each of the measurements used to find the area. Consider a rectangle measured to have sides of length a and b with respective absolute errors Δa and Δb. The relative errors will be $\Delta a/a$ and $\Delta b/b$. Let the area be A, the absolute error in it ΔA, and the relative error $\Delta A/A$ or $\Delta A/ab$. The situation is illustrated in Figure 3–1. The area is ab but may be as small as $(a - \Delta a)$ $(b - \Delta b)$ or as large as $(a + \Delta a)(b + \Delta b)$. The error is equal to the area of the shaded L-shaped strips outside or inside (ab). These differ in area by only the small corner rectangle and, neglecting the effect of this corner, the error areas are $(b\Delta a) + (a\Delta b)$. The area is then $ab \pm (b\Delta a + a\Delta b)$. The quantity in parentheses is the absolute error in the area and it can be written as

$$\Delta A = (b\Delta a + a\Delta b)$$

Divide both sides by the area, A on the left and ab on the right, to obtain the relative error which reduces to

$$\Delta A/A = (\Delta a/a + \Delta b/b)$$

The two quantities in the parentheses are the relative errors in a and in b. In other words, the relative error in the product ab is equal to the sum of the relative errors in a and in b.

Another approach is to consider that the product is actually $(a \pm \Delta a)(b \pm \Delta b)$, which can be written as

$$a\,(1 \pm \Delta a/a)\,b\,(1 \pm \Delta b/b)$$

or

$$ab\,(1 \pm \Delta a/a \pm \Delta b/b \pm \Delta a\Delta b/ab)$$

The last term in the parentheses will be small compared with the others and it can be neglected. The middle terms in the parentheses can be combined to yield

$$A \pm \Delta A = ab\,[1 \pm (\Delta a/a + \Delta b/b)]$$

$$= ab \pm ab\,(\Delta a/a + \Delta b/b)$$

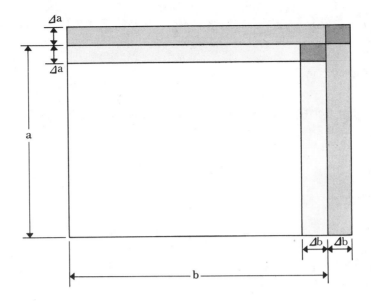

Figure 3-1 An illustration of the error in a product.

So the absolute error in the product is

$$\Delta A = ab \, (\Delta a/a + \Delta b/b)$$

The relative error is $\Delta A/A$, which is, as we have just determined, $\Delta a/a + \Delta b/b$. Again it is seen that the relative errors add when quantities are multiplied.

The equation $\Delta A/A = \Delta a/a + \Delta b/b$ can be multiplied on both sides by 100 to obtain per cent errors. Consequently:

$$100 \, \Delta A/A = 100 \, \Delta a/a + 100 \, \Delta b/b$$

The rule, then, is that *the per cent error in a product is the sum of the per cent errors in the individual numbers.*

POWERS

When a quantity is raised to the nth power it means that the number is to be multiplied by itself n times. For example, $r^2 = r \times r$ and the per cent error in r^2 is equal to the per cent error in r plus the per cent error in r, or two times the per cent error in r. In general, the per cent error in a quantity to the nth power is n times the per cent error in the quantity itself. For instance, the per cent error in r^3 is three times the per cent error in r. If the exponent is a fraction such as a half (a square root), then what!

Let $y = \sqrt{r} = r^{1/2}$

Then $y^2 = r$ or $r = y^2$

The per cent error in r is double the per cent error in y, and it follows that the per cent error in y is just half the per cent error in r. This could be analyzed in a more general way, and the result would be that the rule applies even for fractional exponents. The rule is that *the relative error in a quantity raised to the nth power* is n *times the relative error in the quantity, whether the power is an integer or not.*

DIVISION

Consider the quotient $(a \pm \Delta a)/(b \pm \Delta b)$. a and b may be measured quantities with errors Δa and Δb respectively. This can be written as

$$Q = a\,(1 \pm \Delta a/a)/b\,(1 \pm \Delta b/b)$$

or

$$Q = (a/b)\,(1 \pm \Delta a/a)\,(1 \pm \Delta b/b)^{-1}$$

Expand the last parentheses by the binomial theorem (see Appendix 1) to only the first term in the expansion; then multiply the terms to get

$$Q = (a/b)\,(1 \pm \Delta a/a \pm \Delta b/b \pm \Delta a \Delta b/ab)$$

Neglect the last term, multiply the a/b into the parentheses, and the quotient Q with its error Q becomes

$$Q \pm \Delta Q = a/b \pm (\Delta a/a \pm b/b)\,a/b$$

The last term, the absolute error in Q, amounts to

$$\Delta Q = (\Delta a/a + \Delta b/b)Q$$

The relative error in Q is the absolute error divided by the quantity Q which is just

$$\Delta Q/Q = \Delta a/a + \Delta b/b$$

and this is the sum of the relative error in a and b. We see then that the relative errors add when quantities are divided. Multiplying by 100 would give the per cent errors, and then the conclusion is that *the per cent error in a quotient is equal to the sum of the per cent errors in the numbers used to obtain that quotient.*

In each case, multiplying or dividing, the per cent errors in the quantities add. If a calculation involves both multiplying and dividing, the per cent error in the result is just the sum of the per cent errors in all the quantities involved. For example, if

$$y = \frac{a \cdot b \cdot c \cdot d^n}{e \cdot f \cdot g}$$

the per cent error in y is the sum of the per cent errors in a, b, c, e, f, g and n times the per cent error in d.

The calculation of errors need not be a long, involved process since only one figure accuracy may be used for errors. For instance, the per cent error in 28 ± 1 would be called 4 per cent, not 3.57 per cent.

POSSIBLE AND PROBABLE ERRORS

The error Δa in a measured quantity, a, is an estimate of how much the stated value of that measurement may vary, that is, of its *possible* variation. The actual value may be anywhere in the range $a - \Delta a$ to $a + \Delta a$. If another quantity is represented by $b \pm \Delta b$, the possible error in the sum $a + b$ is $\Delta a + \Delta b$. However, it is improbable that both errors were of the maximum amount and in the same direction; it is possible, but not probable. The error given by the sum of the individual errors indicates the *possible error*, but not the *probable error*. With only two measurements the difference between these two expressions of the error may not be very large, but if many measured quantities are involved in the calculation, the possible error will be unrealistically high.

The previous section on calculation of errors considered only the possible error. There are no apologies for this, because frequently the possible error is what is wanted. It answers the question "just how far off could the calculated answer be?" If the question is "what is the probable uncertainty in the answer?" then the fact that the individual errors may lie anywhere in the range indicated must be taken into account. This is not a simple problem, and it is necessary to go to those who work in statistical analysis for an answer. You should at some time delve into the subject to see what is involved. The result, however, is that if the errors are distributed in a certain way inside the given range, the probable error in the result is the square root of the sum of the squares of the errors in the individual quantities. For example, in adding $(a \pm \Delta a)$ to $(b \pm \Delta b)$, the possible error in the sum is $\pm (\Delta a + \Delta b)$; the probable error in the sum is $\pm \sqrt{(\Delta a)^2 + (\Delta b)^2}$.

If there are many quantities and it is not convenient to represent them by individual letters, they can be represented by a single letter with a subscript. For example, rather than $a \pm \Delta a$, $b \pm \Delta b$, $c \pm \Delta c$, etc., let the measurements be $x_1 \pm \Delta x_1$, $x_2 \pm \Delta x_2$, etc., or, in general, the nth measurement is $x_n \pm \Delta x_n$. The possible error in i values of x is the sum of all Δx_n up to Δx_i. Using the upper case Greek Σ to mean "the sum of," the possible error is

$$\sum_{n=1}^{i} \Delta x_n$$

and the probable error is $\sqrt{\Sigma (\Delta x_n)^2}$, where the range of summation is understood to be over all values of n.

Some numerical values will show the way this works. The first column in Table 3–1 represents a series of fictitious measured quantities that are to be added. In the second column the error in each of the quantities is listed. The sum of the measurements is 201.5; and the sum of the errors in each, which is the maximum possible error in the sum, is ± 1.1. However, the actual errors in the quantities may be randomly distributed inside the error indicated—some of the quantities being too high, others too low. The error in the sum therefore is probably within $\pm \sqrt{0.39}$ or ± 0.6. The sum could be quoted either as 201.5 ± 1.1 (possible error) or as 201.5 ± 0.6 (probable error).

TABLE 3-1   The Sum of Four Numbers Each With an Error

x	Δx	$(\Delta x)^2$
20.5	± 0.2	0.04
10.3	± 0.1	0.01
80.4	± 0.5	0.25
90.3	± 0.3	0.09
$\Sigma x = 201.5$	$\Sigma \Delta x = \pm 1.1$	$\Sigma(\Delta x)^2 = 0.39$

A similar analysis applies to subtraction, multiplication and division. The summarized rules for the calculation of errors follow.

In addition or subtraction:
   The possible absolute error is given by $\Sigma \Delta x_n$.
   The probable absolute error is given by $\sqrt{\Sigma(\Delta x_n)^2}$.

In multiplication or division:
   The possible relative error is given by $\Sigma \left(\dfrac{\Delta x_n}{x_n} \right)$.
   The probable relative error is given by $\sqrt{\Sigma \left(\dfrac{\Delta x_n}{x_n} \right)^2}$.

GENERAL METHODS

The process of finding an error is basically the process of determining how much a function would change if each of the variables (measured quantities) were changed by a certain amount (the errors in the measurements). If the measured quantities are represented by a, b, c, etc., the calculated result is a function of these, $f(a, b, c, \ldots)$. This function changes if the quantity a changes, even if b, c, etc., are perfect. It changes if b changes, even if a, c, etc., are perfect. The process of finding the error involves finding the amount of change in f when each of the measured quantities is varied, and then adding together the changes in f.

The rate of change of a function with respect to one of the variables can be found by differentiation, so it is apparent that calculus methods could be of assistance in the finding of errors. Indeed, in many cases the calculus method is almost the only way to find an error.

In Figure 3-2 is a graph showing how a hypothetical function $f(a, b, c, \ldots)$ varies with the quantity a, assuming b, c, etc., to be constants. A change in a by Δa produces a change in $f(a, b, c, \ldots)$ by Δf. The slope of the function at a is the derivative df/da (keeping all the other quantities b, c, etc., constant). If over the small range Δa, f can be considered to be close to a straight line, the total change, Δf, will be approximated closely by the rate of change, df/da, times the distance Δa.

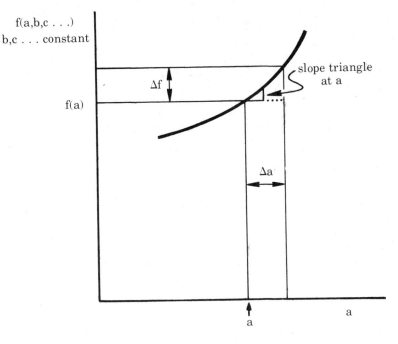

Figure 3-2 The error in $f(a, b, c, \ldots)$ resulting from an error in a of an amount Δa.

The process of differentiating a function of many variables with respect to one variable alone while keeping all the others constant is called partial differentiation, and it is usually indicated by replacing the usual d with a symbol ∂. The error in $f(a, b, c, \ldots)$ due to an error in a is given by

$$(\Delta f)_a = \frac{\partial f}{\partial a} \Delta a$$

By similar reasoning, the error in $f(a, b, c, \ldots)$ due to an error in b is given by

$$(\Delta f)_b = \frac{\partial f}{\partial b} \Delta b,$$

and due to an error in c

$$(\Delta f)_c = \frac{\partial f}{\partial c} \Delta c, \text{ etc.}$$

The total absolute error in $f(a, b, c, \ldots)$ due to the error in all the quantities is

$$\Delta f = \frac{\partial f}{\partial a} \Delta a + \frac{\partial f}{\partial b} \Delta b + \frac{\partial f}{\partial c} \Delta c + \cdots$$

plus variations due to any other variables. The quantities Δf, Δa, Δb, etc., are the errors. Although these are not infinitesimally small, one frequently encounters the notation of calculus.

$$df = \frac{\partial f}{\partial a} da + \frac{\partial f}{\partial b} db + \frac{\partial f}{\partial c} dc + \ldots$$

in which the increments df, da, etc., are treated as errors.

As an example, let $y = ab$. Then the absolute error in y is

$$dy = da\, b + a\, db$$

The relative error in y is $dy/y = dy/ab$. This becomes

$$\frac{dy}{y} = \frac{da}{a} + \frac{db}{b}$$

This is as was previously deduced; when quantities are multiplied, the relative errors add.

There are situations in which this method of determination of error is about the only one that works. If the same quantity appears in two or more places in an equation, the contribution to the total error from that quantity in each of its two positions may tend to cancel. For example, consider that $y = (a + 1)/a$. If a has a measured value of 50, then y will be 1.0200. However, if the true value of a is 51, there having been an error of 2 per cent in the measurement, then the true value of y will be $52/51 = 1.0196$. Calculating errors on the basis of the methods outlined previously, with a 2 per cent error in a, an error in y of 4 per cent would be predicted. Yet the actual error is only 0.04%.

An example of a physical situation in which this occurs is the calculation of the focal length, f, of a lens from measured object and image distances a and b. The focal length is given by

$$f = \frac{ab}{a + b}$$

Both a and b occur in the numerator and denominator, so the errors cannot be calculated by the simple methods described, but partial differentiation must be used. For this particular case, it can be shown that

$$\frac{df}{f^2} = \frac{da}{a^2} + \frac{db}{b^2}$$

The derivation of this is left as a challenge.

As a further example of the use of calculus in finding an error, consider the function

$$y = a \sin \theta$$

What is the error in y due to an error, $\Delta\theta$, in θ and an error, Δa, in a?

$$\Delta y = \frac{\partial y}{\partial a} \Delta a + \frac{\partial y}{\partial \theta} \Delta \theta$$

$$\frac{\partial y}{\partial a} = \sin \theta$$

$$\frac{\partial y}{\partial \theta} = a \cos \theta$$

then $\Delta y = \sin \theta \cdot \Delta a + a \cos \theta \cdot \Delta \theta$.

If θ and a are measured and the errors $\Delta \theta$ and Δa estimated, then the absolute error Δy is easily found. The quantity $\Delta \theta$ must be expressed in the basic unit of radians! Without calculus, how would you do it?

An expression for the relative error, $\Delta y/y$, could also be found. Dividing by $a \sin \theta$ leaves

$$\frac{\Delta y}{y} = \frac{\Delta a}{a} + \cotan \theta \cdot \Delta \theta$$

Again, $\Delta \theta$ must be in radians.

THE ERROR IN A MEAN

It has already been pointed out that there are many natural quantities in which there is inherent variation. A large number of measurements allows calculation of the mean (or average) value. There are two significant quantities which describe the error in the mean. First, there is an expression of the amount of variation present in the set of measurements, and then there is an expression of the accuracy with which the mean is known. The range of variation is basically not a function of the number of measurements, but precision of the knowledge of the mean does depend on the total number of determinations.

The inherent variation can be expressed in two ways, as either mean deviation or standard deviation. The mean deviation is just the average deviation of the individual readings from the average or mean value. This must be computed using absolute values—that is, neglecting the fact that some deviations are positive and some negative.

In symbols, if we let each reading be x and there be n determinations of x, the mean is $\bar{x} = \Sigma x/n$.

The absolute value of the deviation of a reading is $|x - \bar{x}|$ and the mean deviation is

$$\text{m.d.} = \Sigma|x - \bar{x}|/n$$

Consider the set of determinations of the speed of a rifle bullet, each in ft./sec., which are listed in Table 3–2.

TABLE 3-2   Determinations of the Speeds of Rifle Bullets.

Bullet No.	Speed in ft./sec. x	Deviation from Mean $\lvert x - \bar{x} \rvert$
1	1013	13
2	1024	1
3	1009	14
4	1041	18
5	1026	2
Sum	5113	48
Mean	1023	10

The average velocity is 1023 ft./sec. and the average variation of the individual determinations from this is 10 ft./sec. The result of the determinations is that the average velocity of the bullets was 1023 ± 10 ft./sec. (mean deviation).

Other expressions of the variation are the variance and the standard deviation, which are based on rigorous statistical analysis.

The variance is often represented by the square of the lower case Greek letter sigma, σ, and is determined from

$$\text{Variance} = \sigma^2 = \Sigma (x - \bar{x})^2 / n$$

The quantity σ, the square root of the variance, is the standard deviation.

Instead of just taking the sum of the absolute variations from the mean as is done to get a mean deviation, the average of the squares of the deviations from the mean is obtained. This gives the variance, and then the square root is taken to find the standard deviation.

A more rigorous formula for use with experimental data is the estimate, S, of the real standard deviation, σ, in the sample given by

$$S^2 = \Sigma (x - \bar{x})^2 / (n - 1)$$

By this formula, if $n = 1$ then S cannot be determined (0/0); whereas using the formula for σ, if $n = 1$, then the standard deviation is zero. One measurement cannot give any estimate of the deviation, so the formula using $(n - 1)$ in the denominator is obviously more realistic. However, if the number of determinations, n, is large, the difference between S and σ will be small.

There is only a little more work to get a standard deviation than a mean deviation, and most reports of research use the standard deviation. It is, therefore, a process worth learning. Table 3-3 lists a set of numbers, and the method to calculate the standard deviation is illustrated. These numbers are a set of determinations of the counting rate of a Geiger counter when a certain radioactive source is near it. Each value x is the number of counts registered in 10 minutes, divided by ten to obtain the number of counts per minute.

The result of the analysis carried out in Table 3-3 is that the mean of determinations is 3046 counts per minute with a standard deviation of ± 28 counts per minute.

TABLE 3-3 Determinations of the Counting Rate of a Sample of Radioactive Material. The counting rate determined by counting for ten minues is x.

x counts/min.	$x - \bar{x}$	$(x - \bar{x})^2$
3036	10	100
3048	2	4
3057	11	121
3091	45	2025
3068	22	484
3013	33	1089
3054	8	64
3012	34	1156
3010	36	1296
3068	22	484
Sum 30457		6823

Mean: $x = 3046$

$$S^2 = \Sigma(x - \bar{x})^2 / (n - 1)$$
$$= 6823/9$$
$$= 758$$

Estimate of standard deviation:
$$S = \sqrt{758} = 27.5$$

The result, then, is that the mean counting rate is estimated to be 3046 ± 28 counts per minute. This is based on ten determinations; if more measurements were made and included in the calculation of the mean, a different value would be obtained. The larger the number of measurements, the more closely would the mean approach a true, but fictitious, mean of which 3046 is an estimate. The standard deviation ± 28 counts per minute is an indication of the variation in the readings. If an exceedingly large number of readings were taken, the standard deviation, of which ± 28 is an estimate, would approach a value which would enclose 67.5 per cent of the readings. This standard deviation is basically a function of the variation of the readings from each other. It is not an estimate of how accurately the mean has been determined. The accuracy of the estimate of the mean increases (the error in the estimate of the mean decreases) with an increased number of readings. The standard deviation approaches a constant value.

Statistical analysis shows that the accuracy of the mean varies as the square root of the number of determinations used in its calculation. The error in the estimate of the mean, called the standard error, is given by $\pm S/\sqrt{n}$ where S is the estimate of the standard deviation of the sample and n is the number of measurements. In the foregoing example the error in the mean is numerically $28/\sqrt{10} = 9$. The mean counting rate has therefore been determined to be 3046 ± 9 counts per minute or 3046 counts per minute ± 0.3 per cent.

ERROR AND EXPERIMENT DESIGN

Theoretical physicists sometimes predict the occurrence of certain phenomena that had not previously been noticed. The experimentalist then devises a situation or an experiment to determine whether or not that prediction is true. For example, a prediction arising from the theory of relativity is that time depends on velocity. According to this theory, if one clock is kept in a laboratory and another is moved away and then back again, the clock that was moved should have slowed down, and it would be behind the stationary one. If clocks are moved at velocities attainable on earth, however, the effects will not be very great. But it is not enough to buy two wrist watches, leave one in the lab and take the other on a jet aircraft around the world, expecting to find a time difference on return. The magnitude of the expected difference must be calculated beforehand, and the required clock precision found. If available clocks are precise enough, it can be done; but if the calculation shows that they are not sufficiently precise, there is no use in doing the experiment. It is better to have an experiment die in the planning than to use a lot of time, effort and funds performing one that has no chance to succeed.

It has been about 50 years since this "time dilation" effect was predicted, but it is only recently that clocks have attained a precision that makes the experiment feasible. In one experiment, one atomic clock was kept in a lab. Another was flown around the world in an easterly direction (to add to the speed of rotation of the earth) and then in a westerly direction (to subtract from the speed of rotation of the earth). The comparison each time confirmed the prediction of the theory of relativity, but an improvement in precision would be desirable. The experiment depended on calculations of expected errors before it was performed.

In your own laboratory work, the sensitivity of electric meters required for an experiment should be calculated beforehand. If the current in some situation is going to be only a few millionths of an amp, you must be prepared to measure such a small current. This is the case in several of the experiments in this book, although in these experiments the planning has been done, and the proper meters have been specified.

SUMMARY OF THE CALCULATION OF ERRORS

No measured quantity is ever exact. There are errors due to the limits of ability to read the measuring instrument, the calibration of the measuring instrument, and the inherent variations in many physical quantities. The error in a calculated result may be found using the following rules:

1. When quantities are added or subtracted, the absolute errors add.

2. When quantities are multiplied or divided, the relative or per cent errors add.

3. The per cent error in the nth power of a quantity is n times the per cent error in the quantity itself.

4. The probable error is determined by the square root of the sum of the squares of the errors.

5. When a number, n, of determinations of a quantity have been made, each represented by x, then the mean is $\bar{x} = \Sigma\, x/n$.

6. If the deviations from the mean exceed the precision of the measurements, then the error may be expressed in one of the following forms:

The mean deviation, $\Sigma |x - \bar{x}|/n$

The estimate of the standard deviation, $S = \sqrt{(x - \bar{x})^2}/(n - 1)$

The standard error in the mean, $S/\sqrt{n}$

The treatment of experimental errors in this chapter is just an introduction, and for further clarification and instruction the reader is referred to any of a large number of books which have been written on the subject—some of which can undoubtedly be found in any university or college library.

3-1

Rolling Objects

Does the speed with which a rolling object descends an incline depend on the object's shape? Do spheres, solid discs and hollow hoops all roll at the same speed under similar conditions? If the speeds do differ, which object rolls the fastest? This problem can be argued about or solved theoretically; or the ultimate authority, experiment, can be consulted. The time of rolling down an incline can also be compared with the expected time of descent of a frictionless sliding object, which can be found theoretically. Energy loss as a result of rolling friction is usually very small and probably need not be considered. If the objects are all fairly similar, the frictional effects would be similar, anyway.

In this project you are to answer the initial questions experimentally. An inclined track is provided, as well as several objects of the shapes mentioned, perhaps as axles with large wheels. Record the amount of time it takes for each to roll over the same length of track, which should be at least a meter long. A stopwatch accurate to hundredths of a second will be required for the timing. The slope of the incline should be adjusted until the required time of descent is at least two or three seconds.

A convincing result cannot be obtained with only one or two readings of each rolling but the timing should be done several times by each of several observers. Then a mean time for each object to roll along the given path can be obtained. The results for each cannot be said to differ unless the differences exceed the limits of error. It is therefore necessary to calculate the errors for each result and to keep taking data until the results are convincing. You can then list the objects in the order of their speed, from fastest to slowest.

It is also of interest to find the ratio of the time for each to roll down the incline to the theoretical time for a sliding object. This time is calculated using the measured distance, s, along the incline and the height, h, that the incline drops in that distance. In terms of these measurements, the time is given by

$$t = \frac{\sqrt{2\,s}}{\sqrt{gh}}$$

The value of g, 9.81 m/sec^2, is sufficiently accurate, and s and h must then be in meters. The variation in this value of g over the world is only ± 0.2 per cent.

The results should be summarized by listing the objects in order of increasing speed along with the ratio of the rolling time to sliding time. Then a correlation may be found (and commented on) between the speed and the area of concentration of the mass—toward the center or the edge.

Another question—at the discretion of the instructor: Does the speed of a rolling object depend on the size? For a given shape, do small objects roll faster than large ones? The theoretical analysis is possible, and you could investigate it.

Apparatus

Stopwatch — preferably reading to 0.01 sec.

Inclined track, just over a meter long

Steel ball about 4 cm. in diameter

Solid cylinder and hollow cylinder or hoop—each one about 4 cm. in diameter and slightly grooved to stay on the track

Meter scale

Spheres or cylinders of different sizes

3-2

Precision of Resistors

The common type of resistance used in electronic equipment is a mass produced item and is made to be within a certain per cent of the marked value. The value is indicated by three colored bands and a fourth band tells the precision or tolerance. If the fourth band is gold, the actual value should be within 5 per cent of the rated value. A silver band indicates a tolerance of 10 per cent, and no fourth band at all implies that 20 per cent variation is possible. The purpose of this project is to find how much variation from the indicated values actually does occur.

The colored bands which indicate the number of ohms to be expected have the following numerical values:

black	= 0		green	= 5
brown	= 1		blue	= 6
red	= 2		violet	= 7
orange	= 3		gray	= 8
yellow	= 4		white	= 9

The first band gives the first digit; the second band, the second digit; and the third band tells the number of zeros to be added to these digits to give the resistance in ohms.

To measure the resistors with sufficient precision we use a Wheatstone bridge. The slidewire type is very convenient and is described in Chapter Ten. The ohmmeter scale on a multimeter is not adequate for this type of work, nor are the voltage or current scales sufficiently precise for the calculation of the resistance using Ohm's law.

A. Use a Wheatstone bridge to measure at least ten resistors all of the same rated number of ohms, and analyze the results to find the mean and the standard deviation. Also find the difference between the mean of the measured values and the indicated value. Refer to pp. 43 to 45 for the method of calculation.

B. Alternatively you could measure at least ten resistors of different indicated values but of the same tolerance. Find the mean per cent variation from the value given by the colored bands.

Apparatus

Slidewire type Wheatstone bridge
Resistance box
Null detector (galvanometer)
10 resistors of same indicated value and tolerance
10 resistors of different indicated values but the same tolerance

3-3

Order in Chance

Radioactive decay is a random process, for the atoms in a sample of radio-active material are independent of each other. In any given time interval some may decay and some may not, but for a certain sample there will be an average decay rate. A Geiger counter can be used to detect the beta or gamma rays emitted by the decaying atoms, and because the process is random the number of particles counted in a given time by the Geiger counter will differ from one trial to another. To observe this, place a sample of radioactive material near the counter supplied and make some measurements of the number of counts in one minute. You will probably find that they are all different. The question is, how much variation can one expect when the process is a random one such as this? What you find in this experiment will apply in the measurement of any random process.

The amount of variation will be expressed by estimating the variance and the standard deviation in the results. The estimate of the variance is S^2 and given by

$$S^2 = \frac{\Sigma (x - \bar{x})^2}{n - 1} \quad \text{(See page 44)}$$

The individual measurements are designated by x, the number of determinations of x is n and the mean is $\bar{x}$. A relation is to be found between the variance, S^2, or the standard deviation, S, and the average counting rate, $\bar{x}$. The value obtained for S^2 will depend on the number of readings, n. The larger the number of readings, the closer S^2 will be to what you may call the true variance of the quantity being studied.

To find the desired relation, you will make a series of readings of the counts, x, in a constant time interval; calculate the mean $\bar{x}$, then S^2 and S. Take more readings, calculate these quantities with the now larger value of n, and continue until the relation developing between the quantities becomes apparent.

In the calculation of S^2 a little algebra applied at this stage will shorten the work. Expand the term in parentheses in the expression to get

$$S^2 = \frac{\Sigma (x^2 - 2 x\bar{x} + \bar{x}^2)}{n - 1}$$

Sum the terms individually and separate them to get

$$S^2 = \frac{\Sigma x^2}{n - 1} - 2 \bar{x} \frac{\Sigma x}{n - 1} + \frac{\Sigma \bar{x}^2}{n - 1}$$

Multiply each of the first two terms by n/n and recognize that the last term is just n times $\bar{x}^2/(n-1)$ when the summation is carried out. You will see that $n/(n-1)$ occurs in each term so it can be factored out, leaving

$$S^2 = \frac{n}{(n-1)} \left\{ \frac{\Sigma x^2}{n} - \frac{2\,\bar{x}\,\Sigma x}{n} + \bar{x}^2 \right\}$$

The first term in the large parentheses is the mean of the values of x^2, or $\overline{x^2}$. In the second term, $\Sigma x/n$ is just $\bar{x}$ so that term becomes $-2\,\bar{x}^2$ which then combines with the last term to yield $-\bar{x}^2$. The expression for the estimate of the variance then becomes

$$S^2 = \frac{n}{(n-1)} (\overline{x^2} - \bar{x}^2)$$

The term $\overline{x^2}$ is the average of the squares of the values x, and the term $\bar{x}^2$ is the square of the average of x. Are these different? You may investigate this by using the numbers 2, 4, and 6 as values of x and comparing $\overline{x^2}$ to $\bar{x}^2$.

The advantage of this expression for S^2 is that in a table you can list the values of x in one column, list x^2 in another column, and then average each column in finding S^2. In this experiment after ten readings calculate $\bar{x}$, S^2, and S, then continue to 20 readings and use all 20 values to get another estimate of $\bar{x}$, S^2, and S. Repeat the calculations after each ten readings to get values of these quantities for $n = 10, 20, 30$, etc. You will have to use longhand arithmetic or a calculator, but not a slide rule, because the difference between $\overline{x^2}$ and $\bar{x}^2$ is small. You will also have to check and double check your additions and other calculations.

It should soon become apparent to you what the relation is between S^2 or S and $\bar{x}$ for a random process. To assist you, make a table for n, $\bar{x}$, S^2, and S.

To complete the analysis, find what per cent of your readings were within one standard deviation of the mean—that is, between $\bar{x} - S$ and $\bar{x} + S$. Find also the per cent of the readings that were more than $2\,S$ away from the mean. The theoretically expected answers are 67 per cent and 5 per cent.

Some examples will show how to use your results:

1. For a mean rate of 100 counts in a minute, in what range would 67 per cent of the determinations lie?

2. Again for a mean of 100 counts in a minute, what range would encompass 95 per cent of the values of one minute measurements?

3. If you have a mean of 10,000 counts in a given interval, what is the expected standard deviation expressed as a per cent?

Answer all the questions that are on the report sheet.

This project shows, among other things, the large amount of calculation involved in the analysis of data. Yet, rather than putting the data into a computer after the measurements have all been taken, it is recommended that the calculations be carried out after each series of readings. In doing this you will see how the estimated variance changes as the number of readings increases. It goes up and down, but the variations in it become less and less as n increases. S seems to be tending toward some certain value. The mean value, $\bar{x}$, also is different each time, but the variations in $\bar{x}$ decrease as more readings are obtained. These variations become much less than S, so it is apparent that the mean is known to a higher precision than the variance S. This is a good time to turn back and read the section, The Error in a Mean, to see if the information there is consistent with the data gathered in this project.

Apparatus

Geiger counter with timer
Radiation source giving 20 to 30 counts in a 15 sec. time interval

REFERENCES

1. Howlett, L. E.: Announcement on the international yard and pound. Can. J. Phys. *37:*84 (1959).
2. Davies, J. G., and Lovell, A. C. B.: Space research. *Proceedings of the First International Space Science Symposium, 1960.* pp. 515–517.

4

THE LABORATORY RECORD BOOK AND THE REPORT

An experiment is a fleeting thing: It is created, performed, and then exists no longer except in the memory of the experimenter and, more important, as a set of data in a notebook. These cold, hard data are the foundations from which science has grown. It is expected that a similarly constructed experiment would yield a similar set of data and similar conclusions. That this is the case allows us to formulate laws and have faith in them.

The book in which the initial measurements are taken and in which the results are calculated and a conclusion made can be called a *lab record book.* It is hardly possible to emphasize sufficiently the fact that data must be recorded in a permanent form and with enough explanation so that memory is not required to figure out what a set of numbers is all about. You have already recorded the data and results from some experiments in your lab book. Look back at these reports now and judge them from the viewpoint of someone years in the future who wonders what you did and what the results were. After a little thought you will probably realize that only a few sentences and a few words of explanation with the numbers would make it intelligible to such a person. From now on, try to keep your record book with this in mind. The experiment recorded in this form would still be of little value unless the results were communicated to others. An experiment, especially in research, must be made public, so there is a type of written report for this purpose which is often called a *formal report.* This formal report is based on the information that is in the lab record book, and it may take one of many forms. In the case of free research it would be submitted as an article to a scientific journal for printing and distribution. It could take the form of a report to a senior member of a firm, or in your case it will usually be the filling out and submission of the report sheet as directed by the laboratory supervisor. You will probably also be expected to make some formal reports which would be more like a journal article. These reports could include the purpose of the experiment, the method of approach to the problem, diagrams, data, results, a theoretical analysis and conclusions. Sometimes this formal report will be required to be very brief and sometimes very detailed.

An *abstract* of 200 words or less accompanies most journal articles, and to be consistent with the aims of this lab, it should also accompany each formal report. It can be quite an exercise to condense the purpose, main features, and results of an experiment into an abstract.

The laboratory record sheets at the end of the book are almost self-explanatory. The purposes of the record sheets are twofold. First, they will indicate to you whether or not you have obtained the type of data and conclusions that are required. The second purpose is that in the case of large classes the instructor can easily see the data and results for evaluation. Remember, however, that it is basically you who should be satisfied. Mean values and standard deviations can be worked out using the tables on the report sheets. Pertinent graphs should be submitted with the record sheet because the conclusion is often based on the graph. The graphs must be titled, and the axes labeled. Printing should be used on graphs. Slope triangles, if they are used, should be shown. The graphs in Chapter 5 are examples.

Most students prefer to use the laboratory record book for raw data, so it will be described in more detail.

THE LAB RECORD BOOK

While you are performing an experiment, you have to record the data somewhere, and it would save time if it could be recorded directly onto the report sheet. However, sometimes you will want to redo a part of the experiment or check measurements, and then some data may have to be changed. This can lead to messy reports, so obtain a record book for yourself. This can be a coil back or other type of book. Do all the initial recording into this book—*never on loose pieces of paper.* In this book record even the date and the experiment title. Write enough words with each recorded number so that you will know what the number refers to. With the numbers put also a unit; that is, specify whether it is centimeters, millimeters, grams, etc. Neither the neatness nor the style of writing in this book will ever be judged, though its content could be evaluated for completeness. It is difficult to put too much information into the lab record book. More commonly, there will be too little. This record should have almost the status of a personal diary.

An illustration of a portion of a lab record book may help to convey this idea. Suppose we are finding the specific gravity of a quartz crystal by weighing it in air and then in water. The lab record might read like the following, although you would probably do better than this.

s.g. of a quartz crystal—about an inch long.
 wt. in air—23.15 gm.
 in water—9.35 gm.
 wt. lost = 13.80 gm. = wt. of vol. of H_2O
 so s.g. = 23.15/13.80 = 1.675; this is far lower than it should be— try
  again:
 wt. in air—23.15 gm. O.K.

in water—13.50 gm. —had to shorten thread—it hit bottom
loss—9.65 gm.
s.g. = 23.15/9.65 = 2.40—more realistic—wonder if all types of quartz are
the same?

Don't worry about mistakes in the lab record book—that is the place for them. But don't destroy your wrong data either. It is surprising how many times what you at first thought was wrong was later found to be useful data.

If you foresee that you are going to make a series of measurements, make a table for the data in your lab record book. The orderliness of a table saves space and reduces confusion and error. So whenever possible, use tables or columns for recording measurements. Also, work out all the results you can as you go along. In this way, data that are not correct can be spotted immediately and checked.

If your report requires a graph, the table for plotting the graph can be put in the lab record book. In general, you will want to have extra columns for the recording of logarithms of the quantities. Tables of natural logarithms to two places are at the back of this book. Two-place accuracy is generally sufficient for the plotting of graphs.

CONCLUSIONS

The *conclusion* of an experiment is naturally very important, and it is worth saying a few words about conclusions in general.

The word *conclusion* used to describe the findings of an experiment is perhaps unfortunate because it indicates that the experiment is closed, that it can be filed away under the appropriate letter and found when reference is again made to that physical situation. Each experimental conclusion becomes a vital part of the structure of physics. It must harmonize with other physical laws and it may point the way to new experiments or theories. It is the opinion of some scientists that scientific reports should end, not with a conclusion, but with a statement of how that work raises questions which call for further investigation. Suffice it to say that the drawing of a conclusion from an experiment does not mean that the topic is dead and buried, but rather that another portion of the whole of physics has been illuminated and will from then on be a visible part of the whole picture. The illumination may be new to all the physicists of the world or it may be new only to you, the student. In any case, illumination to the world is preceded by an illumination to an individual. You may not always be able to see the wider significance of your experiment, although you should attempt to. At any rate, the experiments which you perform will have been chosen so that they will illuminate the theory, and when you encounter in the theory what you have already investigated in the laboratory, you will understand the importance of the experiments.

The following abstract illustrates these aspects of reporting on experimental work. From an experiment with diamonds a relationship between two quantities was discovered. This result was compared with similar experiments with glass. A speculation was then made about the significance of the result, making the work more valuable to science than it would have been if only the discovered relationship had been reported.

The fracture strength of diamond can be measured by the critical average stress required to produce a ring crack on a diamond surface with a spherical impactor. By using impactors of different sizes and materials, an inverse strength-area relationship has been established for diamond.

This relationship and the similarity between it and that found for glass add to the indications, already reported, of a possible flaw distribution in diamond. Evidence is given that, if such a distribution does exist, it is likely to consist of flaws bigger than point defects.[1]

What could be done next to clarify this a little more?

It is also in the conclusion that extension is made from a law actually found to the probable law which is basic or which would have been found if there were no experimental error. As described in Chapter Two, an exponent near a simple number can be said to probably be a simple number. If a quantity is found to be related to another quantity by a relation $y = kx^{1.05}$ it can be said that probably $y = kx$. If $y = kx^{0.45}$ the true relation is probably $y = k\sqrt{x}$. These generalizations must be made, for they reveal a basic law, even though there may be experimental errors as well as perturbing factors which do not make it exact.

EXERCISE

As outlined, write a formal report on one of the experiments already performed and preface it with an abstract of less than 200 words. The instructor will designate which experiment is to be reported and may wish to give a definite form for the report.

REFERENCES

1. Howes, V. R.: A strength-area relationship for diamond. Bull. Inst. of Phys. and Phys. Soc. *13:*162 (1962).

5

GRAPHICAL ANALYSIS OF EXPERIMENTAL DATA

The "cloak and dagger department" of any government has as one of its major problems the "cracking" of the ciphers or codes of foreign agents. Each time a communication written in a new cipher is encountered, machinery is put in motion to unscramble the message. The methods used are not hit and miss guesses, but organized attacks. The frequencies with which words of certain numbers or letters occur are determined. The frequencies of occurrence of various symbols add another clue. Sometimes these help; sometimes they are of no use, but there is a long list of procedures which, when applied with a goodly measure of intelligence, will probably lead eventually to the "cracking" of the code.

Scientific analysis is similar to a decoding procedure. A message consists of a series of data from experiments. The problem is to find relations among the variable quantities.

A dictate of our analytical methods is that relations can be sought between only two quantities at once. The experiments or the data must be arranged so that all the quantities except two are kept constant, and then the way in which variation in one of these affects the other is studied. Then the procedure is repeated with one of those two variables kept constant and another quantity allowed to vary. This procedure is continued until the situation has been studied under the effect of all possible variables. One of the variables will sometimes be what is called a dependent variable. This quantity depends on each of the others, though the others are independent of one another. In such a case, the independent variables will be dealt with one at a time, and for each one the relation to the dependent variable will be investigated. An example will show what is meant.

Consider the simple pendulum which ideally is a point mass on the end of a weightless string but in the laboratory is approximated by a heavy weight hanging on a light string. The time required for one swing of the pendulum, which is called the period, T, could depend on the length, l, of the pendulum; the mass, m, of the bob; and the angle θ through which it swings. The period, T, is the dependent variable, since it may vary with each of the quantities: l, m, and θ. The independent variables are treated one at a time in a series of experiments. First, for instance, l and m are kept constant and the period, T, is measured for different angles of swing, θ. Then l and θ are kept constant, but the mass of the bob is varied. Finally, with the effects of m and θ constant, l is varied. Then when the three sets of data are analyzed, the manner in which the period of the pendulum varies with each of those quantities (the mass of the bob, the angular amplitude of the swing, and the length of the pendulum) is determined. After such an experiment, the experimenter will feel that he knows something about a simple pendulum. Such a series of experiments would lead to a formula $T = c\sqrt{l}$ where c is constant. Neither m nor θ would appear (if θ was fairly small). The constant would, after theoretical analysis, be found to be given $2\pi/\sqrt{g}$ where g is the acceleration caused by gravity.

The analysis of data from an experiment is always reduced to finding the relation between just two quantities, so methods used to determine various types of relations will be discussed.

If two quantities are related by some regular function, for each value of one quantity, there is a certain value of the other. If one varies by a small amount, the other will also vary by some small amount, and a continuous variation in one quantity leads to a continuous variation in the other. The outcome is that when a graph is plotted of one quantity against the other, a line is formed. So the first use of a graph is to show if a relation exists, for if it does, the values of one when plotted against the values of the other will appear to fall on a line. In order to define a line, many points are required. The number required depends on the shape of the line; but if its shape is unknown, the larger the number of points, the better. Two points define a line only if it is known that the line is straight; otherwise a larger number of points is required.

In Figure 5-1A the melting points of various elements are plotted against their specific gravity. No simple curve could be drawn to go through the points, so there is apparently no correlation between the melting point of a substance and its specific gravity. Neither low melting point nor high melting point is associated with high density. In Figure 5-1B the velocity of sound at 0°C. in various gases is plotted as a function of the molecular weights of the gases. These points appear to fall along a curve of the type shown. Apparently the velocity of sound in a gas does depend on the molecular weight of the gas. This graph shows only that the relation exists. The next step in the analysis is to find the type of relation or, in different words, the form of the equation of the line which fits the points. The form of the lines illustrating several different formulae is illustrated in Figure 5-2, and these are only a select few of the curves that could be drawn. The curves shown are limited to just a few powers and roots. By examination of an unknown curve, it is impossible to tell what the equation describing it is, unless that curve is a straight line. The straight line is the clue to graphical analysis. It and it alone can be identified. The problem is then to find a method to plot the experimental data

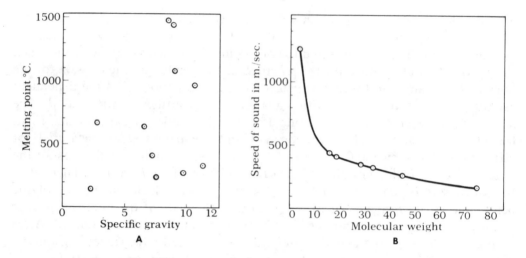

Figure 5-1 Graphs can show whether or not two quantities are related. **A** is a plot of melting point vs. specific gravity. These two quantities are apparently not related. On the other hand, graph **B**, a plot of the velocity of sound against molecular weight for gases, shows that a relation between these two quantities exists.

to give a straight line. There are several ways in which this can be accomplished. Some ways apply only to special cases, and some are more general. There is no completely general method, and in tackling an unknown, several trials may have to be made before the solution is obtained.

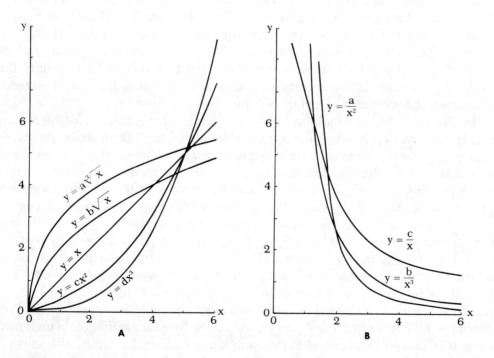

Figure 5-2 Graphs illustrating various mathematical relations. Only the straight line can be identified by inspection.

LINEAR OR FIRST POWER RELATIONS

The general equation of a straight line is of the form

$$y = a + bx$$

a is the value of y when $x = 0$, and is sometimes called the y-intercept. If the line goes through the origin, then $a = 0$. The quantity b is called the slope of the line and is commonly expressed as rise/run. The rise and run are measured in the units indicated on the y- and x-axes respectively. With this understanding of slope, the values of y must be plotted in the vertical direction, and those of x must be plotted horizontally. If the slope b is zero, then y does not depend on x.

Some experimental data, will, of course, give a straight line when one quantity is plotted against the other. In Table 5-1 are data relating the velocity of a falling ball to the time after an arbitrary $t = 0$. These data are presented graphically in Figure 5-3. The velocity, v, is on the y-axis, and the time, t, is on the x-axis. Note how the axes are labeled, how only a few principal values are marked on each axis, and how each experimental point is marked boldly as a dot with a circle around it. These points do not lie exactly on a straight line; we do not expect them to, because experimental data are never exact. The points do *suggest* a straight line, however, so a straight line has been drawn among the points. Such a line is called the "best fit" straight line. It goes above some points and below others, but it is an honest attempt to show the trend of the data. The data do not suggest that a smooth curve would fit the points better than such a straight line. The equation of the straight line could be represented in this case by

$$v = v_0 + at$$

since v has been plotted on the y-axis, and t has been plotted on the x-axis. The value of v when $t = 0$ is v_0 and the slope of the line is represented by a. On the best

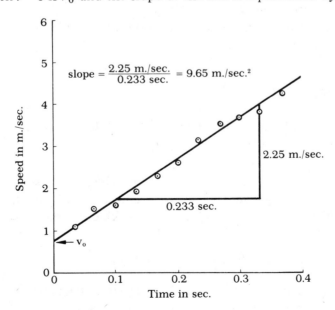

Figure 5-3 A graph of the speed of a falling object plotted against time.

fit straight line a slope triangle has been drawn and the rise and run of each has been labeled in terms of the units shown on the respective axes. The slope, a, is then calculated to be 9.65 meters/sec.^{2}. Note the inclusion of the units with the calculated slope.

TABLE 5-1   Some Determinations of the Speed of a Falling Object.

Time in sec. (x)	Velocity in meters/sec. (y)
0.033	1.08
0.067	1.50
0.100	1.64
0.133	1.96
0.167	2.34
0.200	2.66
0.233	3.11
0.267	3.48
0.300	3.66
0.333	3.84
0.367	4.27

These results may be then compared with the results expected from theoretical analysis. The fact that the experimental points indicated a straight line showed that the object in that experiment fell with constant acceleration, at least within the limits of that experimental method. The value of the slope, which was the acceleration, was within about 1.6 per cent of the acceleration due to gravity accepted for the location. Whether or not this is sufficiently close to the accepted value to allow it to be said that other forces on the falling object were negligible can be determined only if the numerical uncertainty, or error, in the experimental value is calculated.

POWER LAWS

A very common type of relation is one of the form

$$u = kv^n$$

where u and v are the variables, and the power to which v is raised can be integral, fractional, positive, or negative; k represents a constant. This equation includes all those represented by the curves in Figure 5-2. Taking the log of both sides of the above equation gives

$$\log u = \log k + n \log v$$

Now let $y = \log u$, $x = \log v$, $b = \log k$, and the equation becomes

$$y = b + nx$$

which is the equation of a straight line. The quantity $\log u$ can be plotted in the y direction, and $\log v$ in the x direction. A straight line would then result, the slope of the line being the exponent n and the constant k being the antilog of the y-intercept.

This method can be illustrated with an actual set of data. Table 5–2 lists the four planets nearest the sun; their mean distances from the sun in meters, R; and the number of years it takes each to go once around the sun, which is called the period T. Shown also are the logarithms of R and T. Figure 5–4 is a graph of T against R. The points appear to lie on a curved line. Figure 5–5 is a graph of $\log T$ against $\log R$. The points fall accurately along a straight line. The slope of the line, calculated from the slope triangle shown, is 1.50. The equation relating the period of a planet to its mean distance from the sun is therefore one of the form

$$T = \text{const. } R^{1.50}$$

Squaring both sides gives

$$T^2 = KR^{3.00}$$

where K represents a new value of the constant. Dividing by $R^{3.00}$ shows that the ratio $T^2/R^{3.00}$ is constant. This relation is known as Kepler's third law, which was first made public in 1619 after analysis of experimental data and was later shown to follow from the law of gravity postulated by Newton.

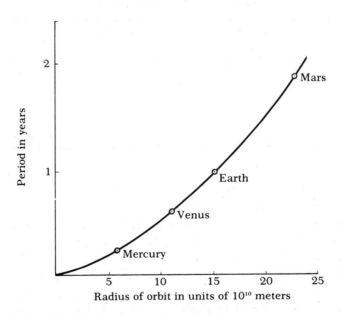

Figure 5-4 A graph of the periods of planets against the radius of their orbits. The smooth curve that results shows that a relation exists between these quantities.

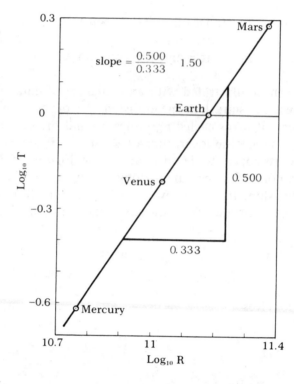

Figure 5-5 A graph of the log of the periods of planets against the log of the radius of their orbits. The straight line that results shows that the relation between these quantities is in the form of a power law. The slope of the line is 1.50, so the relation can be expressed by $T = \text{constant} \times R^{1.50}$.

TABLE 5-2 For the Inner Four Planets the Mean Orbital Radius, The Periods, and the Logarithms of These Quantities are Listed.

Planet	Mean Distance in Meters (R)	Period in Years (T)	Log R	Log T
Mercury	5.79×10^{10}	0.241	10.7627	−0.6180
Venus	10.81×10^{10}	0.617	11.0383	−0.2097
Earth	14.95×10^{10}	1.000	11.1746	0.0000
Mars	22.78×10^{10}	1.881	11.3576	0.2744

It is left to you to determine the value of the constant K in the form of the equation $T^2 = KR^3$ applied to the planets.

The log-log plot just described is a general method by which an unknown exponent can be determined. If the exponent in the relation is theoretically predicted, another type of graph may be used to test the prediction. This may also be illustrated with the data about planetary orbits as shown in Table 5–2. The equation $T^2 = KR^3$ can be compared to the equation of a straight line through the origin, $y = Kx$, if in the y-direction are plotted the values T^2 and in the x-direction the values R^3. The result would be a straight line of slope K. If the data were seen to lie on a straight line, the relation would be verified for the planets. The graph is shown in Figure 5-6. The theory is seen to describe the actual situation within very close limits.

In general, if the relation is expected to be of the form $u^n = cv^m$, put $y = u^n$ and $x = v^m$. Then the graph of u^n versus v^m or y versus x would yield a straight line if the theory adequately described the physical situation. Such graphs are useful only if theoretical analysis suggests what to plot. In order to analyze for an unknown relation, the log-log plot should be used.

The expected equation may be more complicated. For instance, the equation relating the distance, s, traveled in a time, t, by an object having constant acceleration, a, is

$$s = v_0 t + 1/2 \, at^2$$

v_0 is the value of the velocity at $t = 0$. Unfortunately, neither plots of s against t nor against t^2 nor a graph of log s against log t would yield a straight line, and as we have seen, only a straight line can be identified with any degree of certainty. However, dividing the equation through by t gives

$$s/t = v_0 + 1/2 \, at$$

Now, if s/t is plotted in the y-direction and t in the x-direction, the result would be expected to be a straight line of slope $a/2$ and y-intercept v_0. Both unknowns can be determined from such a graph.

There could be many other situations, of course, and it is often necessary for the student or the researcher to use his ingenuity to devise a plot which would yield a straight line and which would therefore lend itself to analysis.

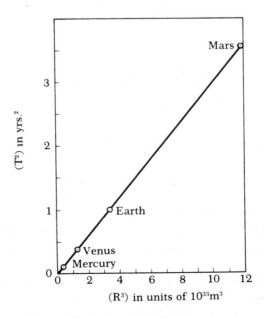

Figure 5-6 A graph of the squares of the periods of planets against the cube of the radii of their orbits. The straight line shows that the relation between these quantities is described by $T^2 = \text{constant} \times R^3$.

The use of the type of plot based on a theoretical analysis is not recommended, however, because it prejudges the answer. It would be easy to get almost a straight line and accept the result. It is much wiser to use a log-log plot and find what the exponent *really* is. Then small variations from the theoretical value would be apparent and these variations may be of great significance for further theoretical analysis.

EXPONENTIAL OR LOGARITHMIC RELATIONS

Some physical situations are described by an exponential or a logarithmic relation. These relations can be expressed in powers of either 10 or e and in logs to the base 10 or to the base e. The quantity e is a natural, irrational number, 2.71828 . . . Another irrational number with which you are familiar is π, which is 3.14159 . . . and which also cannot be represented by any simple fraction. Logarithms to the base e are called natural or Napierian logarithms, often written $\log_e$ or ln. Just as the logarithm to the base 10 of a number is the power to which 10 must be raised to give that number, so the logarithm to the base e of a number is the power to which e (2.718 . . .) must be raised to give the number in question.

Since $10 = e^{2.303}$, it can be readily shown that natural logs are related to logs to the base 10 by

$$\log_e v = 2.303 \log_{10} v \text{ or } \log_{10} v = 0.4343 \log_e v$$

There is a short table of natural logarithms at the end of the book.

An exponential relation between the quantities u and v is one which is described by an equation of the form

$$u = e^{kv}$$

Taking the logarithm to the base e of both sides of the equation yields

$$\log_e u = kv$$

Converting to the common logarithms to the base 10

$$\log_{10} u = 0.4343 \, kv$$

Now, if data which are related by such an exponential relation were to be analyzed, a graph with either $\log_e u$ or $\log_{10} u$ in the y-direction and v in the x-direction would result in a straight line of slope either k or 0.4343 k, depending on the base used for the logarithms.

Exponential relations describe many natural phenomena. Among these are growth and decay processes and absorption phenomena. The data shown in Table 5–3 illustrate such a relation. These data consist of measurements of the width of successive growth spirals of a particular sea shell *(Catapulus voluto)* illustrated in Figure 5–7. In Figure 5–8 some graphical plots of the data are illustrated. In part A the width, w, of the spiral is plotted against the spiral number, n. The result, a

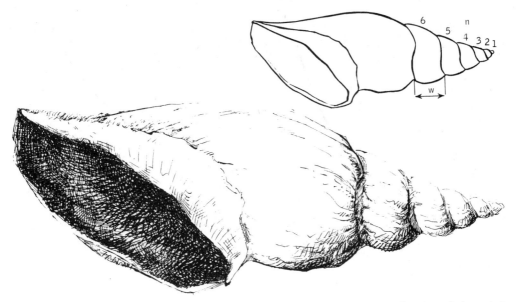

Figure 5-7 The sea shell *Catapulus voluto*. The regular increase in the size of the spirals indicates a possible mathematical relation between the width and the number of the spiral. (Drawing prepared by Miss J. Grey.)

smooth curve, not a straight line, tells only that the relation is not described by an equation of the form $w = kn$. Part **B** of Figure 5-8 is a plot of $\log_{10} w$ versus $\log_{10} n$ and again the data show a curved line, so there is not a power law relation. The third graph, of $\log_e w$ against n, shows a straight line so the relation is apparently of the form

$$\log_e w = \log_e a + kn \text{ or } w = ae^{kn}$$

The constant k is the slope, which in this case is 0.503, and the intercept is 0.43, the antilog$_e$ of which is 1.54. The equation relating w to n is then

$$w = 1.54\, e^{0.503\, n}$$

The solid curve of Figure 5-8A was plotted from this formula. If the relation was of the form $w = k \log n$, then a plot of w versus $\log n$ would have been required to obtain a straight line.

TABLE 5-3 The Width of Successive Spirals of a Sea Shell of the Type *Catapulus Voluto*. The number of the spirals is n and the width in millimeters is w.

n	w	$\log_{10} n$	$\log_{10} w$	$\log_e w$
1	2.5	0.000	0.398	0.92
2	4.5	0.301	0.653	1.50
3	6.5	0.477	0.813	1.87
4	11.5	0.602	1.061	2.44
5	20.0	0.699	1.301	3.00
6	31.0	0.778	1.491	3.43

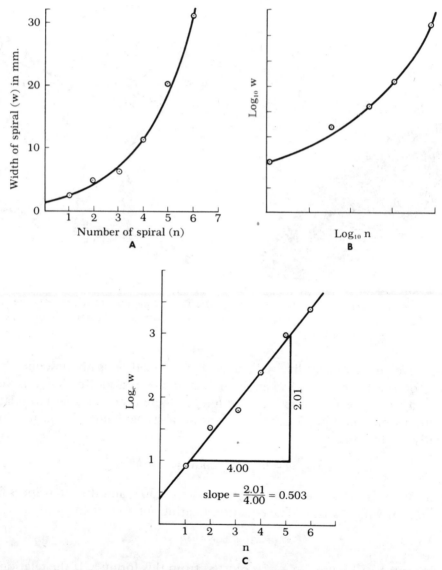

Figure 5-8 In **A** the widths of the spirals of the shell shown in Figure 5-7 are plotted against the number of the spiral. A relation of some form is apparent. In **B** the logarithm of the width is plotted against the logarithm of the number, and the curve shows that the relation is not in the form of a power law. In **C** the natural logarithm of the width is plotted against the number of the spiral. The straight line shows that the relation is exponential and is described by $w = 1.54e^{0.503n}$.

Before jumping to conclusions about sea shells in general, we should consider the words of S. Earnshaw. He said, "Numerical results are no certain test of a theory when limited to a few cases only."[1] Do we have enough evidence to generalize?

TESTING A THEORETICAL RELATION

A theoretical relation can be tested without resorting to a graphical plot which yields a straight line. A graph may be plotted of the expected results based on the application of the theory to the physical situation. Then the experimental points can be shown on the same graph. If the theory adequately described the situation, the experimental points would lie along the theoretical line, being no farther away from it than their experimental error. If the points do not follow the theoretical line, then the theory is inadequate. The result of an experiment analyzed in this fashion is illustrated in Figure 5-9. The data are measurements of the intensity of the earth's magnetic field at various altitudes. The solid line on the graph was calculated on the basis of the inverse cube law and the points shown were from measurements in a rocket-borne magnetometer. The inverse cube law adequately describes the magnetic field to an altitude of 100 km., but above that altitude the theory does not describe the actual situation. Why not? Nature does not easily yield her secrets, and it is this reluctance on her part that keeps theoretical and experimental physicists intrigued! This figure also illustrates why scientists are not easily satisfied. A rocket reaching an altitude of only 100 km. would have found no discrepancy from theory, but it would not have validated the theory for all altitudes. Uncharted regions are full of surprises.

If there are two theories which predict different results, then two theoretical curves can be shown on the graph and the experimental points may lie along one of the lines, thereby showing which theory is preferred.

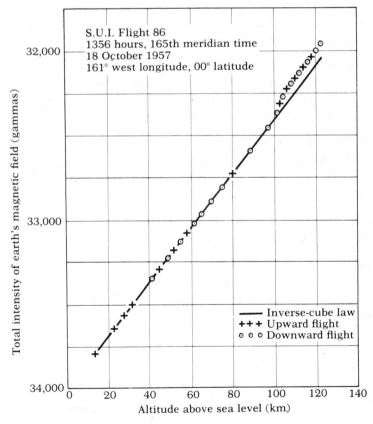

Figure 5-9 The earth's magnetic field as a function of altitude. The measurements shown by the points were taken with a rocket-borne magnetometer. The solid line is based on an inverse cube law which is shown to hold to an altitude of only 100 km. (Modified from Dieminger, W.: *In* Benson, O. O., Jr., and Strughold, H., eds.: Physics and Medicine of the Atmosphere and Space. New York, John Wiley & Sons, 1960, p. 94.)

Special types of graph paper are available to facilitate the plotting of the logarithms of quantities. These papers are not marked in uniform squares, but the distances along the axes are made proportional to the logarithms of the numbers which are shown on the axes. The spacings are similar to the spacings of the numbers on a slide rule. If one axis is marked logarithmically and one uniformly, the paper is described as *semilog paper* and is used to search for or investigate exponential relations. Another kind of paper is divided logarithmically on both axes and is called *log-log paper*. This paper is used to test for power type relations, the geometrical slope of a line on such paper being the exponent relating the variables.

SUMMARY

Graphical analysis is seen to be a very powerful tool to search for relations between measured physical quantities. The only curve which can be identified with any degree of certainty is a straight line, and the efforts are directed to finding a plot which will yield a straight line. Consider two quantities, u and v.

1. If a graph of u against v gives a random scatter of points, u is not related to v at all.

2. If a graph of u against v yields a straight line, the relation is of the form $u = a + bv$ where a is the intercept and b the slope. If $b = 0$, u does not depend on v.

3. If the graph of u against v is not a straight line, a plot of log u against log v will test for a power law relation. If a straight line is obtained with such a logarithmic plot, the relation is described by an equation of the form $u = av^n$ where n is the slope of the line and a is the antilogarithm of the intercept.

4. If the relation is of an exponential form, $u = ae^{bv}$, a plot of the logarithm of one quantity against the other directly would yield a straight line. It may be necessary to try the log of one quantity against the second and then the log of the second against the first.

5. In some special cases theory may suggest a special plot which may be made and will yield a straight line if the theory applies.

6. A straight line is not necessary to test a theoretical relation. A graph can be constructed based on the theory and then the experimental points can be shown on the same graph. If the experimental points follow the theoretical line, the theory is then shown to describe the situation.

Problems

1. Complete the following table and then graph each of the following functions against x: $y = \sqrt{x}$, $y = x^2$, and $y = 1/x$.

x	$\sqrt{x}$	x^2	$1/x$
0	0	0	1
1	1	1	2
2	1.414	4	0.50
4	2.00	16	0.25
6			
8			
10			

2. Complete the following table in order to become familiar with natural logs.

x	$\log_{10} x$	$\log_e x$
0.5		
1.0		
1.5		
2.0		
5.0		
10		
20		
100		

3. The following sets of numbers represent data collected in experiments. Find in each case the equation relating the two sets of figures. Use graphical analysis, as described in the text, to find the form of the equation and then the constants in the equation. Because the data are experimental, a perfect fit to a straight line may not be obtained in any example.

(a)

y	x
0	0
0.37	0.13
0.71	0.25
2.68	0.95
9.59	3.50
13.63	4.82

(b)

y	x
94	307
106	327
155	408
231	535
460	917

(c)

y	x
1.39	1.2
1.95	2.4
2.76	4.8
3.96	10.0
7.00	30.0

(d)

y	x
7.13	2.45
8.38	2.08
5.97	2.91
3.44	5.03
4.22	4.50

(e)

y	x
1.66	0.4
3.02	1.6
5.80	2.9
8.65	3.7
21.2	5.5
45.0	7.0

4. The following figures give, for a certain location, the dose rate from the fallout from a nuclear weapon as a function of time after the explosion. The fallout ended before the first measurement was made. Find the equation relating dose rate and time for this mixture of fission products. Logs to the base 10 may be used for the graphs.

Time in Hours	Dose Rate in Roentgens Per Hour
10	63
20	27
40	12
100	4.0
200	1.7
400	0.75
1000	0.25

5. Figure 5-1B shows that there is a relation between the velocity of sound in a gas and the molecular weight of the gas. Table 5-4 lists some representative gases with their molecular weights and the velocity of sound in each of these gases. From these data find the relation between these quantities. There will be deviations from a line because the velocity depends in a small way on the structure of the molecules. These differences should be more noticeable on the straight line graph than on a curve.

TABLE 5-4   Velocity of Sound in Gases.

Gas	Molecular Wt.	Velocity of Sound at 0°C. in ft./sec.
NH_3	17	1361
CO	28	1106
CO_2	44	846
CS_2	75	606
H_2	2	4165
CH_4	16	1417
NO	30	1066
NO_2	46	859
O_2	32	1041

EXPERIMENT

5-1

Newton's Rings

This experiment deals with a phenomenon which is not usually part of everyone's experience. To observe the phenomenon, place the lens supplied (it has a very long focal length) on top of the piece of plate glass. Look down on this system with a light source approximately behind you and you should see colored rings. They will stand out better if a black background is used. These rings are referred to as Newton's rings, having been described by him in his book *Opticks* in 1675. Figure 5-10 is a reproduction of a part of a page of the fourth edition of *Opticks* (1730) in which the rings are described by Newton as he saw them formed from white light. Figure 5-11 is a reproduction from *Opticks* of Newton's own drawing of the rings.

> *Obf.* 4. To obferve more nicely the order of the Colours which arofe out of the white Circles as the Rays became lefs and lefs inclined to the Plate of Air; I took two Object-glaffes, the one a Plano-convex for a fourteen Foot Telefcope, and the other a large double Convex for one of about fifty Foot; and upon this, laying the other with its plane fide downwards, I preffed them flowly together, to make the Colours fucceffively emerge in the middle of the Circles, and then flowly lifted the upper Glafs from the lower to make them fucceffively vanifh again in the fame place. The Colour, which by preffing the Glaffes together, emerged laft in the middle of the other Colours, would upon its firft appearance look like a Circle of a Colour almoft uniform from the circumference to the center, and by compreffing the Glaffes ftill more, grow continually broader until a new Colour emerged in its center, and thereby it became a Ring encompaffing that new Colour.

Figure 5-10 Newton's own description of what we now call Newton's rings. They are described as having been formed from white light. This is an excerpt from Newton's book *Opticks*.

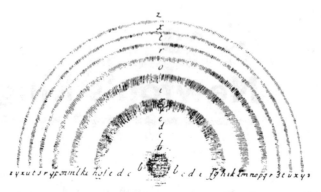

Figure 5-11 Newton's own drawing of Newton's rings as it appeared in his book *Opticks.*

The rings become closer and closer together with increasing number from the center. It is the purpose of this investigation to experimentally find the formula that relates the radius of the ring to its number from the center. A traveling microscope is to be used to measure the diameter of a large number of rings. The rings are to be produced with monochromatic light (from a sodium vapor lamp or mercury lamp with a green filter). To achieve vertical illumination, a sloping glass plate is to be placed to reflect the light downwards (Fig. 5-12). The slope is adjusted until the rings are seen brightly when you look vertically downward onto the reflecting plate. The light adjustment can be made with the microscope tube moved aside; and after bright rings are seen as you look straight down, the microscope tube is moved into place to measure them. There is no need to measure all the rings, but measure a variety of sizes—from small to as large as you can conveniently count. In measuring the radius of a ring read the microscope position when it is set on the left side of the ring, then when it is moved to the right side of the same ring—as indicated in the report sheet. This is most easily done by counting about 30 to 50 rings to the left and recording the microscope positions on a select few of the rings as the microscope is moved across to the right. Record the data in the table and use graphical analysis to relate r to n.

In your results give your actual formula and generalize as to the probable relation between r and n. What factors do you think may affect the sizes of the rings? As food for thought, what causes the rings to appear using such a simple arrangement?

To have those questions answered, look up "Newton's rings," "Interference," or "thin films" in the index of your text.

You could also use a white light source to produce the rings, just as was done by Newton.

Apparatus

Measuring microscope
Plano-convex or double convex lens with a focal length of 3 meters or more
1 piece of ordinary glass about 1 × 4 inches
1 piece of plate glass about 1 × 4 inches
Black card or black cloth on a board
Sodium lamp or mercury lamp with green filter
Stands and lens holders (as necessary)

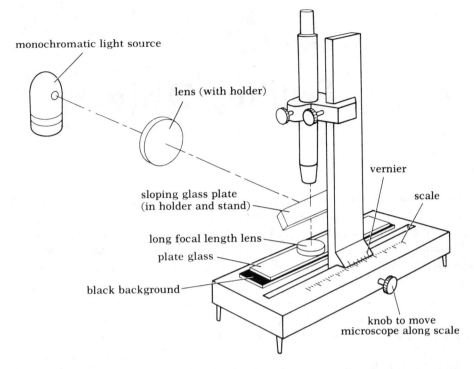

monochromatic light source

lens (with holder)

vernier

scale

sloping glass plate
(in holder and stand)

long focal length lens

plate glass

black background

knob to move
microscope along scale

Figure 5-12 The arrangement of the apparatus to produce and to measure Newton's rings.

5-2

Falling Objects

The purpose of this experiment is to use precision apparatus to find how the velocity of a falling object changes with time.

A falling steel ball is to be photographed using a Polaroid Land camera and a stroboscope. The room is to be dark except for the stroboscope illumination. Using the strobe lamp we will make a multiple exposure photograph showing the position of the ball at evenly spaced time intervals. The distance between successive image positions is the distance traveled in the time between flashes of the lamp, and from this the average velocity over the time interval can be found. From a series of such velocity calculations the relation between velocity and time can be found graphically. With these small time intervals, the average velocity over the interval can be taken as the instantaneous velocity at the *time* halfway through the interval.

The apparatus consists of a stand to hold the camera and a release mechanism for the ball. Examine this mechanism to see how it works. Attached to it is a portion of a meter scale which will appear in the photograph and which will be used to convert from distances measured on the photograph to actual distances in space. A measuring microscope which measures to 0.01 mm. is used to measure the positions of the images of the ball on the film.

To perform the experiment place two ball bearings on the release mechanisms (so that the photograph can be cut to give two students a record), set camera to **B** (at this position the shutter will be open as long as the shutter release is held down), and set the stroboscope flashing at 2400 flashes per minute. You are now ready. Darken the room and in fairly quick succession open the camera shutter, release the balls, and close the camera shutter. You may then turn on the room lights and develop the film according to the instructions furnished with the camera.

Place the photograph under the microscope and read and record the position of each of the dots representing the positions of the ball. Measure with the microscope between two marks on the image of the meter stick to find a conversion factor from microscope scale readings to distances in space. All data should be put into your record book in tabular form. Suggested column headings are: dot number, time, microscope reading, distance between dots on film, distance in space, velocity, and distance from first dot.

Analyze the results graphically according to one of the procedures outlined on page 61 or page 65.

Answer the following questions:

1. Did the ball have a constant acceleration?
2. What was the acceleration?

3. What was the velocity at the time which you call $t = 0$?

4. Do the equations given on pages 61 and 65 satisfactorily describe the motion of that ball or did air resistance have a significant effect?

Apparatus

Stroboscope
Polaroid Land camera
Release mechanism for dropping balls
Ball bearings
Measuring microscope

5-3

Trajectory

The object of this experiment is to make some investigations regarding the velocity of a particle in a trajectory.

A small ball is to be thrown at an angle and its path recorded by photographing the position at regular intervals by means of a Polaroid Land camera and a stroboscope. A suitable arrangement of apparatus is shown in Figure 5–13. The room is darkened and the stroboscope is adjusted to produce 2400 flashes per minute. Next, the camera shutter is opened, the ball is thrown, and the camera shutter is then closed and the picture developed. On the film is recorded the position of the ball each fortieth of a second. The experiment consists of taking the picture and then analyzing the trajectory shown.

The velocity of a particle in a trajectory can be resolved into horizontal and vertical components which will be referred to as v_x and v_y respectively. These components are found from the horizontal displacement, Δx, and vertical displacement, Δy, which occur in a small time interval, Δt. Then $v_x = \Delta x/\Delta t$ and $v_y = \Delta y/\Delta t$. These quantities are illustrated in Figure 5–14.

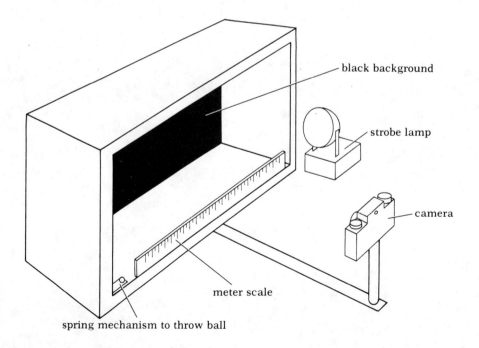

Figure 5-13 The arrangement of the apparatus to photograph a steel ball moving in a trajectory. Stroboscopic illumination is used.

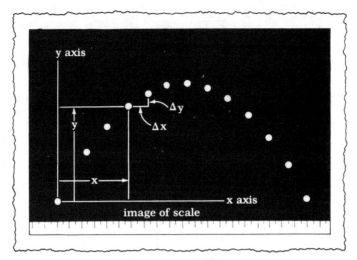

Figure 5-14 A drawing made from a stroboscopic photograph of a trajectory. The coordinate positions x and y are illustrated, as are Δx and Δy which represent the x and y displacements in the interval between flashes of the lamp.

A horizontal line can be shown on the picture by the edge of a leveled meter stick, which will also serve to indicate the scale of the picture.

For each dot measure the x and y positions using either a divider and ruler (estimating to at least the nearest tenth of a millimeter) or a traveling microscope. The measuring microscope shown in Figure 5-15 with the microscope tube moved to a horizontal position is especially suitable for this experiment. The trajectory photograph can be taped to a vertical mount (the side of the microscope box), but take care to level it properly. Then the readings of the horizontal and vertical microscope scales may be considered to be the x and y positions on the film. Determine a scale factor by measuring along the image of the meter scale. Convert each of the x and y measurements to distances in space from the first measured dot and tabulate these. Find each Δx and Δy, which are the differences between successive recorded positions of the ball, and find also the x and y velocities in each interval. Call the time of the first velocity obtained $t = 0$ and put in the table the time, t, for each v_x and v_y.

Make two graphs—one of x versus t and y versus t on the same sheet and the other of v_x versus t and v_y versus t again on the same sheet. Because velocities downward are negative, on the latter graph the x-axis should be drawn at about the center of the sheet. Determine the equations for whatever straight lines appear and fill in the report sheet.

Apparatus

    Apparatus for throwing a small ball in a trajectory
    Stroboscope
    Polaroid Land camera
    Measuring microscope or dividers and steel scale
    Meter stick
    Level
    Steel balls

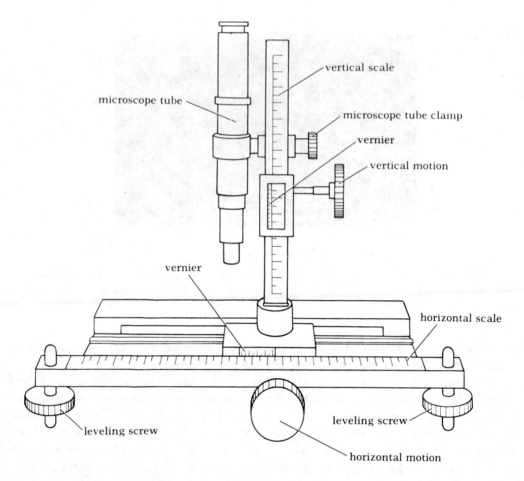

Figure 5–15 A measuring microscope. This type has scales in two directions and the microscope tube may be positioned from the vertical to the horizontal.

5-4

Moment of Inertia of a Solid Disc

In the study of rotary motion the quantity called *moment of inertia* occurs in the same way that mass or inertia does in linear motion. Mass or inertia is given by the ratio of force to acceleration (Newton's second law), and similarly moment of inertia, I, is given by the ratio of torque, L, to angular acceleration, α.

$$I = L/\alpha$$

The moment of inertia depends not only on the mass but on the radius of rotation of the mass. In this project the way in which the moment of inertia of a solid disc varies with the radius will be found. In order to find this, a number of discs of the same mass but different radii are supplied. For each a torque is applied, the angular acceleration is measured, and I is calculated. Then, by graphical analysis I is found as a function of r, and the mass is related to one of the constants in the equation.

To carry out the experiment, measure the radius, R, and the mass, M, of each disc. Tabulate the data. Then fasten each disc in turn to the axle as shown in Figure 5-16. Wind the string around the cylinder and attach an appropriate weight (mass m) to the string. Allow it to fall, spinning the disc. This mass may be different for different discs: For each disc choose a mass that will result in an accurately measurable time to fall. From the measured distance and time of fall of the weights, calculate the linear acceleration of the falling weight. Then using the radius of the cylinder, calculate the angular acceleration of the disc. Calculate it again with no disc attached in order to find the moment of inertia of the axle alone, I_0, which must be subtracted from each moment of inertia found with a disc on the axle.

The analysis in more detail is as follows:

1. The linear acceleration of the falling mass m is easily found using $s = 1/2\ at^2$ because $v_0 = 0$. The quantities s (the distance the weights fall) and t are measured, and a is calculated.

2. The angular acceleration α is related to the linear acceleration a by $\alpha = a/r$. r is the radius of the cylinder around which the string is wound plus half the thickness of the string.

3. The torque applied is the tension in the string supporting the weight times the radius at which it acts. This tension, T, is found as follows (see Fig. 5-17A): The net downward force on the falling mass is $mg - T$ and this is equal to ma by

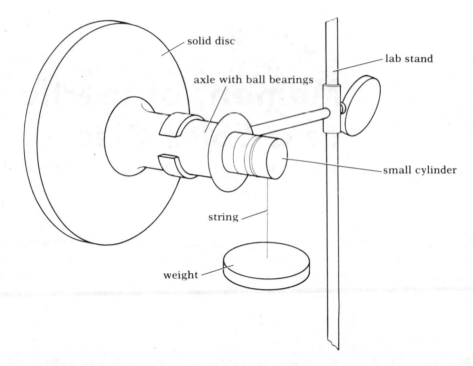

Figure 5-16 The apparatus used to measure the moment of inertia of a disc. The hub of a bicycle wheel can be used for the axle.

Newton's second law. So $mg - T = ma$, or $T = m(g - a)$. m and a are determined and the tension T is calculated. The gravitational acceleration, g, may be taken as 9.81 meters/sec.^{2} or as given by the instructor.

4. The torque is given by $L = Tr$ (see Fig. 5-17**B**) where r is the radius previously described under number 2.

5. The moment of inertia is then L/α, which is really that of the disc plus the axle, or $I + I_0$.

Tabulate all data as requested on the report sheet and do the graphical analysis, finding all the constants in the formula relating the moment of inertia to the radius R of the disc. Speculate on the probable general form, including how the mass of the disc enters the formula.

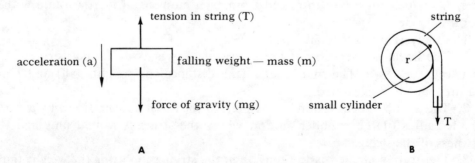

Figure 5-17 Diagrams used to find the moment of inertia of a disc. Diagram **A** shows the forces on the falling mass. The method to calculate the torque is shown in **B**. The distance r is to the center of the string.

Apparatus

Set of 4 or 5 discs of equal mass but different radii—adapted to fit on the axle
Bicycle axle with added small cylinder
Stand and right angle clamp
String
Weight holder and weights
Stopwatch
Meter scale

5-5

Tension in a Cord on a Cylinder

The object of this experiment is to determine the ratio of the tensions (T_0/T) on the two ends of a cord in contact with a cylinder as a function of the angle of contact, θ between the cylinder and the cord.

The apparatus consists of a vertically mounted board on which is fastened a small cylinder as is illustrated in Figure 5-18. Tangential to the cylinder are lines indicating angles from 0 to 2π radians every $\pi/4$ radians. A weight of about 1500 gm. is hung on a cord which is put in contact with the cylinder and is supported by means of a spring scale in such a way that the 1500 gm. weight moves downward at constant speed. The tension T_0 is 1500 gm. of force. The tension T is indicated on the spring scale. Values of T are to be obtained for angles of

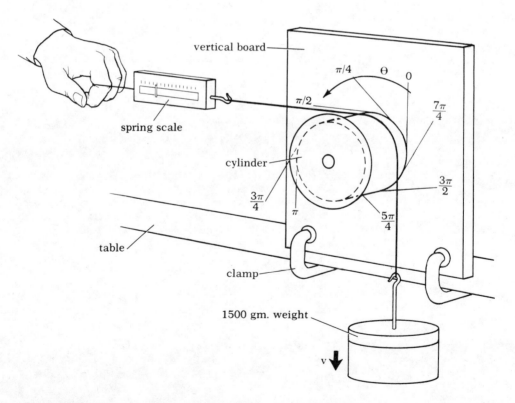

Figure 5-18 The apparatus used to study the relation between the tension on the two ends of a cord in contact with a cylinder.

contact from 0 to at least 4π radians at intervals of $\pi/4$ radians. Measurements can be made beyond $\theta = 4\pi$, but T becomes so small that the precision of measurement becomes low.

Tabulate all data as they are obtained.

Using graphical analysis, determine the relation between T_0/T and θ, the angle of contact between the string and cylinder. Comment on the effectiveness of one man holding against a strong pull by snubbing a rope with a post.

The theoretical analysis of this situation can be approached in the following way:

Consider a segment of the string, Δs, at angular distance θ from the beginning of contact with the cylinder as in Figure 5-19A. At the distance θ the tension in the string is T, and at $\theta + \Delta \theta$ it is $T + \Delta T$. If the motion of the string is in the direction of T_0, then at $\theta + \Delta \theta$ the tension will be less than at θ because of friction, and ΔT will be negative. The segment of the string can be considered to be in equilibrium if the speed is constant and if the centripetal acceleration is negligible. This means that the forces on the segment balance, and considering them as in Figure 5-19B, where ΔF is the frictional force, we can write $T + \Delta T + \Delta F = T$, or $\Delta T = -\Delta F$.

The frictional force is given by μN, where N is the normal force between the segment and the cylinder and μ is a coefficient of friction. The normal force arises because the tensions on the two ends of the element do not act in the same straight line because there is an angle, $\Delta \theta$, between them. From Figure 5-19C we can see that $N = T\Delta \theta$. These relations, $N = T\Delta \theta$, $\Delta F = \mu N$, and $\Delta T = -\Delta F$, can be combined to give $\Delta T = -\mu T \Delta \theta$. If the quantities ΔT and $\Delta \theta$ are allowed to approach the limits dT and $d\theta$, this expression can be integrated to give the tension in the string as a function of the angle θ. The constant of integration is found from what we could call the boundary condition; that is, when $\theta = 0$, the tension is T_0.

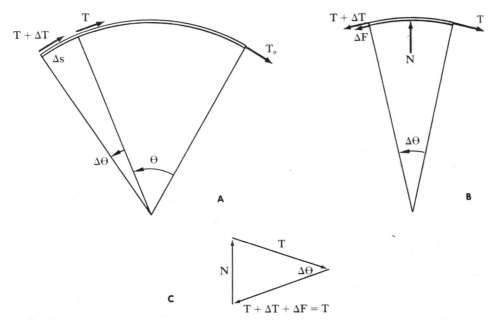

Figure 5-19 Diagrams for the analysis of forces resulting from the contact of a cord with a cylinder.

This integration is left for you to carry out and then you can compare your experimental result with your theoretical result. It may require some manipulation to express them in the same form. You will then see how the coefficient of friction enters the expression, and from the experimental result you will find the coefficient of friction between the string and the cylinder.

Apparatus

Vertically mounted board—about 60 cm. across, with a cylinder about 10 cm. in diameter in the center and with tangential lines as intervals of $\pi/4$ radians

Cord (fishline)

Weight holder with weights to total 1500 gm.

Spring scale reading to 1500 gm.

5-6

Fluid Flow in Tubes

The rate at which a fluid flows through a tube or pipe depends on such factors as the pressure forcing it along, the length and radius of the tube, and the viscosity of the fluid. This project is to find experimentally how the rate of flow varies with the radius of the tube. All other factors will be kept constant.

One could speculate that if the radius of the tube is doubled, and the area is therefore increased four times, the rate of flow would increase four times. Expressed in another way, since the area of cross section depends on the square of the radius, perhaps the rate of flow will also vary as the square of the radius. But in the larger tube the fraction of the fluid which is near the wall is less than in the smaller tube, so perhaps frictional or viscosity effects will not be as large. Your problem is to find experimentally what the relation really is.

A series of glass tubes all of the same length but of different diameters is provided. They are supported horizontally and are connected in turn by rubber tubing to a constant level water tank which is made as shown in Figure 5-20 and which is designed to provide water at a constant pressure. The constant level tank is mounted on a small stand and its height is adjusted to give a rate of flow which

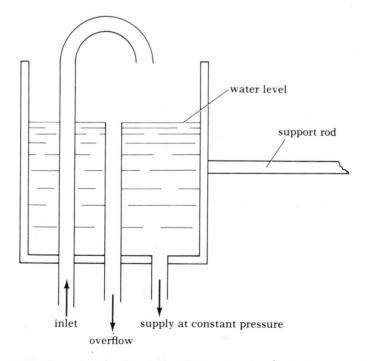

Figure 5-20 A constant level tank to deliver water at a constant pressure head.

will allow about 250 ml. to flow through the largest tube in about a minute. The water supplied should be adjusted so that there is some overflow from the constant level tank.

Use the beaker to collect water flowing through the tubes and measure the time required to collect at least 100 ml. from each tube. Determine the amount of water collected by weighing it.

Measure in this way the rate of flow through each tube.

Use a measuring microscope to determine the radius of each tube.

Analyze the data graphically to find the realtion between the rate of flow and the radius. (Express it as an equation.) Give also the probable basic relation.

The flow of a fluid through a pipe can be analyzed theoretically and you could find the analysis in the library or perhaps from your instructor. Do not do this, however, until your experimental result has been obtained.

Apparatus

Set of 4 tubes from 0.5 mm. to 2 mm. in diameter and 50 cm. long
Constant level water tank
Small lab stand
Rubber or plastic tubing
250 ml. beaker
Balance
Stopwatch
Measuring microscope

5-7

Light from a Linear Source

It is well known that with a point source the amount of light falling on a given area is inversely proportional to the square of the distance from the source. The question is, if the source is not effectively a point, but of large dimensions, does the inverse square law hold, or does some other law? This experiment will be concerned with the way in which light decreases with the distance from a long linear source.

Suitable light sources are several fluorescent tubes placed end to end or a series of small bulbs spaced about 4 to 6 inches apart. A total length of 6 to 8 feet is adequate for measurements up to 2 feet from the source.

The light can be measured with a barrier-type photocell or a solar battery. Unfortunately, there is considerable reflection from the surface of such cells if the light strikes at an angle. Therefore, the indicated reading does not properly include the light from distant parts of the linear source. The photocell provided has one half of a cylindrical plastic rod cemented over it. With such an arrangement the variation in sensitivity with direction is less than if only the flat cell is used.

The main part of the experiment consists of the measurement of the photocell current at different distances from the light source. But the magnitude of the current from the photocell is not necessarily proportional to the intensity of the light falling on it. Therefore, it is necessary to calibrate the cell. To do this, a light bulb with a small filament is provided, and for this the inverse square law can be assumed to hold. The photocell current at some arbitrary distance, d_0, can be measured, and the light intensity falling on it called E_0. The inverse square law is used to calculate the intensity, E/E_0, at other distances, and a graph of E/E_0 is plotted against photocell current i. This is used to convert the readings near the linear source into light intensity readings.

One reason for measuring the linear source first is so that the calibration can be done over the same range of readings as those for the project.

Graphical analysis is to be used to determine the relation between light intensity on the photocell and the perpendicular distance from it to the linear light source. This relation will, of course, be in the form of an equation. Give not only your actual result but also your guess as to the probable basic form of the relation.

The situation can also be analyzed theoretically. To do this we accept the inverse square law to describe the variation in intensity of light from a point source. The linear source will be considered a series of small sources which can be treated as points.

Let the intensity per unit length of the source be I. Then the intensity of a portion which has a length dl (as shown in Figure 5-21A) is $I \times dl$. At the detector the light flux from this element of the source is given by $dE = I dl/r^2$. The distance r can be expressed as $r = x/\cos\theta$, and the length dl as $dl = r d\theta/\cos\theta$.

It will be left to you to substitute these into the expression for dE, to integrate along the whole light source (from θ_1 to θ_2, where θ_1 will have a negative value), and then to find the expression for E in terms of x when the length of the source is infinite.

Compare your experimental and theoretical results.

The method just described is not for the illumination of a surface normal to a perpendicular line from the source. For oblique rays, at angle θ from the normal, the illumination is given by $I dl \cos\theta/r^2$. The $\cos\theta$ term takes into account the fact that the oblique light spreads over a larger area. Analyze this situation using the same method as before to see if the expected illumination would follow the same law.

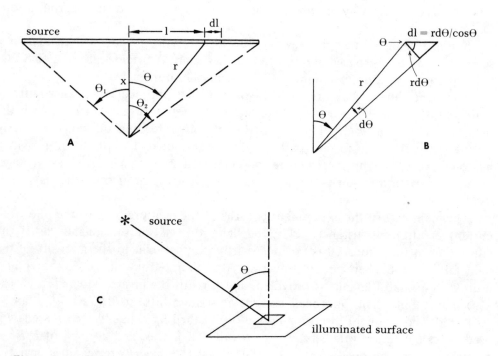

Figure 5-21 Diagrams to assist in the theoretical development of the expression for the intensity of light near a long linear source.

Apparatus

For calibration
Optical bench
Light source with small filament
Photocell
Microammeter

For light from a linear source
  Linear light source
  Photocell adapted as suggested
  Microammeter
  Meter scale

5-8

Field Near a
Magnetic Dipole

A short bar magnet is a magnetic dipole, having two opposite poles close together. You have learned that an inverse square law describes the variation of the field with distance from a single pole, but what happens when two poles of different polarity are close together? The object of this experiment is to find a law which describes the variation of field strength with distance from a dipole. The experiment can be most easily done either in the line joining the poles or perpendicular to that line.

Instead of actually measuring the field, it is sufficient to determine at each distance how the field from the magnet compares with the earth's field. This can be obtained by arranging the apparatus so that the field of the magnet is perpendicular to the direction of the field of the earth. A compass will then show the direction of the resultant of the two, and the ratio of the field from the magnet, B_m, to the horizontal component of the earth's field, B_e, is given by $B_m/B_e = \tan \theta$, where θ is the angle by which the compass is deflected from its normal position in the field of the earth. (See Fig. 5-22.) In this way the field being measured is said to be a certain fraction or multiple of the earth's field. The arrangement of compass and magnet for measuring the field perpendicular to the axis of the magnet is shown in Figure 5-23A. Figure 5-23B shows the arrangement for measuring the field along the axis of the magnet. The procedure will be described only for the situation shown in Figure 5-23A.

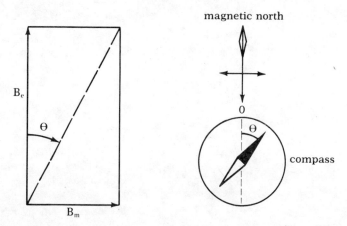

Figure 5-22 The use of a magnetic compass and a vector diagram to find the strength of a magnetic field, B_m, in terms of the horizontal component of the earth's field, B_e.

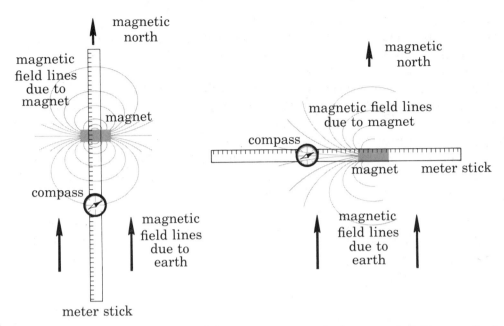

Figure 5-23 Two arrangements of the apparatus to study the magnetic field near a dipole. In **A**, the arrangement is for studying the field perpendicular to the axis of the dipole and in **B**, for studying the field along the axis of the dipole. In each case, the field from the magnet is perpendicular to the field due to the earth at the position of the compass.

To measure the magnetic field in the way suggested above, the field from the magnet must be perpendicular to the field from the earth. The pattern of the magnetic field about the bar magnet is indicated in Figure 5-23, and in the two situations shown (measuring the field along the axis of the magnetic dipole or perpendicular to it) the field of the magnet at the position of the compass is perpendicular to the earth's field. Before beginning the measurements, the meter stick must be properly oriented with the aid of the compass and with the magnet far removed (not near another student's apparatus, however).

The compass should be set on two wooden or plastic meter sticks (not steel) with their zero ends together under the center of the compass. With the magnet far removed, the two meter sticks are to be lined up along the magnetic meridian with the aid of the compass. The compass should then be turned to zero on its scale. Both ends of the needle should be read and if both ends do not indicate zero (or if one does not indicate zero and the other 180 degrees—depending on the marking), the compass should be turned so both ends are an equal amount off the desired mark. Some precise compasses have a short magnet which lines up along the magnetic lines of force and a long pointer perpendicular to the magnet.

The short magnet is then set across the meter stick and the deflection of the needle noted. Both ends should be read and the readings should be recorded in the table. The mean deflection is then obtained. The distance from the center of the magnet to the center of the compass should also be recorded in the table. A column is allowed for $\tan \theta$ which is B_m / B_e. Readings should be taken for at least ten positions over as wide a range of distance as it is possible to obtain a satisfactory deflection.

For maximum reliability of results, another set of observations should be obtained with the magnet on the other side of the compass. Both sets should be plotted on one graph, but use two types of points—perhaps circles and x's. Various graphs should be plotted until a straight line graph is obtained, and then the equation relating B_m/B_e to x should be found.

An extension of the experiment is to find the answer to the following questions: At the same distance from the center of the dipole, is the strength of the magnetic field along the axis the same as that perpendicular to it? If it is not the same, how are the field strengths in those two directions related? If time allows, find the answers experimentally!

To analyze the dipole theoretically, consider the magnet as consisting of two poles of strength m, a distance l apart. (In electricity there is an analogous situation of the electric dipole, which consists of two equal sized charges of opposite sign, separated by a distance l.)

Magnetic field can be considered as the force on a unit N pole. The field strength at a distance x from a pole of strength m is then given by m/x^2; that is, an inverse square law holds.

Consider the field along the axis of a dipole as in Figure 5-24.

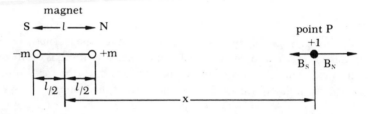

Figure 5-24 A diagram to assist in the theoretical analysis for the strength of the field along the axis of a dipole.

At the point **P** the field due to the **N** pole is $B_N = m/(x - l/2)^2$ and that due to the **S** pole is $B_s = -m/(x + l/2)^2$. We could also write these as

$$B_N = \frac{m}{x^2(1 - l/2x)^2} \quad \text{and} \quad B_s = \frac{-m}{x^2(1 + l/2x)^2}$$

The total field is the sum (considering the sign). To simplify the resulting expression, expand the parentheses using a binomial expansion, and consider that l/x is small compared with 1. Compare the result of this analysis with the experimental result.

In the expression for the field you find the quantity ml. This is called the magnetic moment, M, and, one can speak of the magnetic moment of a magnet or a current-carrying coil, or of a proton or an electron without the concept of a magnetic pole at all.

The equation describing the field perpendicular to the axis of the dipole is left as an exercise.

Apparatus

Short, strong bar magnet
Compass
Meter sticks

5-9

A Capacitor in an A.C. Circuit

A capacitor, or condenser as it is sometimes called, consists of two conducting plates separated by an insulator (a dielectric). Electrical connections are made to the plates, but there is no connection between them. Because of this, it seems at first that they would be of no value in an electric circuit. However, an example of a condenser is the series of intermeshing plates used to tune a radio receiver to different stations. Air is the dielectric between the plates, and the capacity is varied by changing the portion of the plates which is together. Another type of capacitor has two strips of metal foil separated by thin plastic sheets and wound into a cylindrical form. The unit used to measure capacity is the farad, a capacitor of one farad holding a charge of one coulomb on each plate (+ and −) when charged to a voltage of one volt. The most commonly used units are actually the microfarad (μF) (10^{-6} farads) and the picofarad (pF) (10^{-12} farads).

A capacitor in a D.C. (direct current) circuit acts almost like a break in the circuit. Once the condenser is charged, no current flows. In an A.C. (alternating current) circuit the situation is different. The capacitor charges in one direction and when the voltage source changes polarity, the capacitor discharges and then charges in the opposite direction. This continues so an alternating current *will* flow in a circuit having capacity. When you begin to work with A.C. circuits, including those with audio and radio frequency currents, you will find that the important circuit elements are resistors, capacitors, and inductors or coils. This project is an introduction to the behavior of capacitors. Some find it amazing that they are used in electric circuits, for there is no electrical connection between the two leads from the capacitor.

Just as the current in a D.C. circuit depends on the size of the resistance, the current in an A.C. circuit depends on the size of the capacitor. In a D.C. circuit we refer to the value of the resistance as found from Ohm's law given by the ratio of voltage to current, or

$$R = V/I$$

where V is the voltage in volts, I is the current in amperes, or amps, and R is the resistance in ohms. For an alternating voltage V applied across a capacitor, an alternating current I will flow. The quantity associated with the capacity and analogous to resistance is called the reactance of the capacity (X_c) and is given by the ratio V/I. That is,

$$X_c = V/I$$

Again, with V in volts and I in amps, the reactance X_c is in ohms. (Be careful to change your milliamp readings to amps for the calculation.)

We might speculate that the reactance of a capacitor would depend on its size and perhaps also on how quickly the current changes; that is, the frequency. The purpose of this project is to find how the reactance varies with capacity. The frequency will be kept constant. The procedure will be to connect different capacitors across the A.C. line, to measure the line voltage and the current to the capacitor, and then to calculate the reactance. The equation relating the reactance to the capacity can then be found using graphical analysis. Find also the numerical constants in the equation for comparison later to theoretical values.

If an A.C. source of variable frequency is available, find how reactance varies with frequency when the capacity is constant.

The circuit for the project is shown in Figure 5–25. The project is designed to make direct use of a 120-volt A.C. source such as the power provided for homes and laboratories in North America. In some parts of Europe and in Great Britain the power is supplied at 220 volts. Care should be taken not to touch any bare wires or connections when the apparatus is plugged in. Disconnect it from the voltage source before making any changes. An uncomfortable shock will occur at 120 volts, and a dangerous one at 220 volts, if two places are touched simultaneously with both hands. For this reason a physicist frequently works with one hand either behind his back or in his pocket. It is a distinguishing characteristic of those in the profession.

It has been said that there is no electrical connection between the two leads from the capacitor. To find its resistance to a direct current, use a D.C. source provided instead of the A.C. wall outlet and a D.C. voltmeter and milliammeter. Measure V and I, and calculate the resistance using Ohm's law.

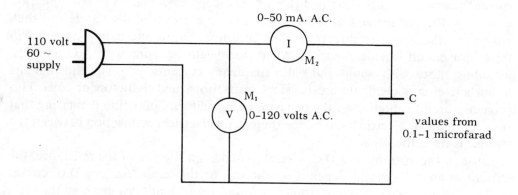

Figure 5-25 The circuit used to study the effect of a capacitor in an A.C. circuit.

Apparatus

110 volt, 60 cycle outlet and cord
A.C. voltmeter–0–120 volts
A.C. milliammeter–0–50 mA
Capacitor substitution box or 5 capacitors ranging from 0.1 to 1 μF

For additional work
    audio oscillator
    D.C. voltage source, 10 or 100 volts
    D.C. voltmeter, appropriate for source
    D.C. milliammeter

5-10

Resonance, Capacity, and Inductance

In many physical situations a phenomenon known as resonant vibrations may occur. An air column may resonate at a certain frequency (Experiment 6-4) and use is made of this phenomenon in organs and various other musical instruments. In fact, almost all musical instruments are based on a form of resonance. Resonance can also occur in electric circuits which have coils and condensers (inductance and capacity). To observe this, connect a coil and condenser in series across an audio oscillator which is a source of voltage of variable frequency. Put a voltmeter across either the coil or the condenser as shown in Figure 5-26, and watch it as the frequency is varied. You will find one frequency at which the meter reading is a maximum. This is the resonant frequency. If you change the value of the condenser (capacitance) or the coil (inductance), the resonant frequency changes. An example of the use of this effect is in the selection of a station by a radio. Each radio station broadcasts with a certain carrier frequency, and to receive that station the radio must be tuned to resonate at that frequency. Different stations may be selected by varying either the capacity or the inductance of a resonant circuit. Which one is varied depends on the receiver—although most commonly it is the capacitor, a set of interleaving metal plates.

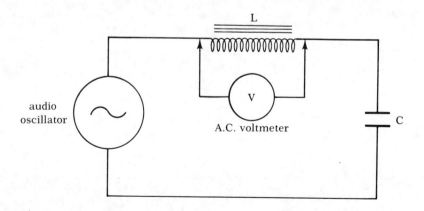

Figure 5-26 The circuit used to study resonance in circuits which have capacitors and inductors.

The object of this experiment is to find the relation between the resonant frequency and the sizes of the capacitor and inductance.

The apparatus consists of an audio oscillator, a variety of inductances, and several capacitors which are connected in turn as in Figure 5–26. Across the inductance is a high resistance A.C. voltmeter used to detect resonance. The experiment will be in two parts: First the inductance will be kept constant and the value of the capacitor will be varied. Then with a constant value of capacity, the inductance will be varied. Finally, an equation will be found in which the resonant frequency, f, is expressed as a function of inductance L and capacity C.

To investigate the relation with capacity, suitable values for the components are a 10 henry inductance, six values of capacitors between 0.0001 μF and 0.01 μF (100 pF to 1000 pF), and an oscillator which can be varied over the audio frequency range. The meter should be about 5000 ohms per volt, A.C., with 6 and 30 volt ranges.

For a given value of capacity the voltage across the inductance will depend on the frequency setting of the oscillator. At resonance, when the impedance is a minimum, the current reaches a maximum, so the voltmeter across the inductance will show a maximum. Find the resonant frequency for each value of capacity. Tabulate the observations. Plot graphs until a straight line is obtained, and then find the equation relating the resonant frequency to the capacity. What does the relation actually seem to be?

To investigate the relation with inductance, use a 0.001 μF capacitor and ranges of inductance from about 2 to 150 henries. Carry out the experiment just as for the previous part and find the equation relating f to L. The voltmeter will have to be placed across the capacitor.

Combine the two equations to write f as a function of L and C. Look up the relation in a text *after* you have completed your analysis. Compare the constants in your empirical formula with the constants theoretically expected.

Apparatus

Audio oscillator
Inductances–about 2, 10, 30, 60, and 150 henries
Capacitors–0.0001, 0.0002, 0.0005, 0.001, 0.002, 0.005, and 0.01 μF
A.C. voltmeter (multimeter or V.T.V.M.)
Connecting wires

The ranges of values of inductors and capacitors are only suggestions. If inductances of that maximum size are not readily available, try to obtain larger capacitors than suggested in order to compensate. Part of the laboratory process is to try variations!

5-11

The Law of Growth and Decay

There are processes in nature for which the rate of change of the size of the quantity is proportional to the size at that time. If the process is one in which the change in each time interval is positive, it is a growth process or if the change is negative, it is a decay process. There is a general expression or law that describes such processes and it is your object in this experiment to find the law of growth and decay.

The law will hold only in the situations in which there is a direct relation between the rate of change and the size. Consider the case in which there are N rabbits. The number of new rabbits, ΔN, which arrive in a time interval, Δt, will be proportional to N, at least if N is sufficiently large. When you have 10,000 rabbits, you expect 10 times as many new ones per day as you did when you had only 1000 of the animals. Expressed mathematically, $\Delta N/\Delta t = KN$ where N is a proportionality constant which you find from measured values of ΔN, Δt, and N.

The only type of process that we are dealing with is one which is described by an equation of the form $\Delta x/\Delta t = Kx$, where x is the size of the quantity. K will be positive for growth processes and negative for decay processes. So we want to find some physical situations which are described by this relation and then find the equation that relates x to t. This will be a very general equation. Several possible experiments will be described; you should do at least two of these to see in which diverse phenomena the same law holds. The instructor may, of course, assign more.

A. COOLING OF A WARM OBJECT

It was Newton who first suggested that the rate at which an object cools would be proportional to the temperature difference between it and its surroundings. That is, where x is the temperature above the room temperature, the rate of decrease of temperature would be proportional to x, or $\Delta x/\Delta t = -Kx$.

You will be provided with a small container with a holder and two thermometers. Put warm water (about 10°C. above room temperature) into the container and place it on the holder. Insert the thermometer, holding it loosely with a clamp so that it won't fall out and get broken. The arrangement is shown in Figure 5-27. Read the thermometer in the water and the other one on the table every two minutes for about 20 minutes and, of course, tabulate the

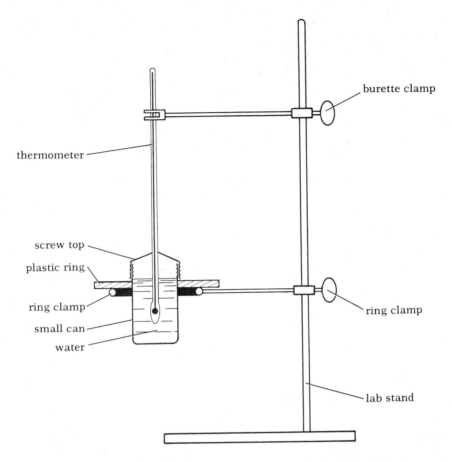

Figure 5-27 The arrangement of the apparatus to find the law describing the cooling of a warm object.

readings. Calculate the temperature excess, x, and $\Delta x/\Delta t$ for each $\Delta t = 2$ minutes. Graph $\Delta x/\Delta t$ versus x to see if the desired relation holds, and if it does, find K. The temperature differences are small, so the error in each value of Δx is large. Do not be discouraged by erratic points, but does the relation seem to be good? Then graphically find how x is related to t and find where the K occurs in this equation. For the presentation of your results, give your actual equations and also speculate on the general relation.

Apparatus

2 thermometers to 0.1°C.
Small container with hole in lid (a 35 mm. film can may be used)
Stand with holder for container and clamp for thermometer
Clock or watch

B. THE DECAYING OSCILLATION

A swinging pendulum gradually comes to rest, largely because of the viscosity of the air through which it swings. Now, as the oscillations become smaller, the speed of the bob in each part of the swing decreases and the resistance effect is proportionately less. If the speed is sufficiently low, then resistance is proportional to velocity and we may expect that the amplitude of the swings decreases at a rate proportional to their magnitude.

To perform this experiment use a light bob on a long pendulum swinging in front of a meter scale. Read the equilibrium position, set it swinging, and note the amplitude every five swings until it effectively stops. Tabulate all data. Then, where A is the amplitude and the number of swings between readings which is proportional to Δt is called Δn, plot $\Delta A/\Delta n$ against A to see if the desired relation holds. You will probably find it convenient to add columns for ΔA and $\Delta A/\Delta n$ to your table. If you are reasonably satisfied that it does indeed apply, find K. Then plot A against n, find the relation between A and n, and also see where the K enters this equation.

Apparatus

Large stand and pendulum clamp
2 meters of nontwisted cord such as fishline
Light bob such as a ping pong ball
Meter scale

C. THE ABSORPTION OF LIGHT

When light shines through an absorbing medium, its intensity gradually decreases—but in what manner? This is not a problem of decay in time; rather, one might say, of decaying intensity with distance. The intensity will follow the law that we are looking for if the change in intensity, ΔE, when the light passes through a thickness, Δx, is proportional to the intensity, E, falling on that absorbing thickness. That is to say, if $\Delta E/\Delta x = -KE$, it is in this class of phenomena. Rearranging this equation, you get $\Delta E/E = K\Delta x$. This shows that for a given thickness, Δx, the fraction of the radiation absorbed, $\Delta E/E$, will be constant. K is called the absorption coefficient and is the fraction absorbed per unit thickness. For this to hold, the thickness Δx must be very thin.

The steps in the experiment are to show that the phenomenon follows the basic relation that $\Delta E/\Delta x = -KE$ and to find the equation relating E to x. The light intensities will be measured with a photocell, so read the part of Chapter Ten that deals with the type of photocell that you have. The current output from the cell is not necessarily proportional to the light intensity falling on it, so it must be calibrated.

First run the absorption curve. Set up the apparatus as in Figure 5–28. The lens just above the lamp focuses the light in a parallel beam through the solution, and the second lens concentrates the light onto the photocell. The lens should be placed to make the light spread out fairly evenly over the whole surface of the cell. The filter should be of a color that is complementary to the color of the absorbing solution. Why?

With no absorbing solution in the container, call the light intensity on the photocell E_0. The photocell gives a current, I_0, which should be near full scale on the microammeter setting of your meter. Pour in solution to a depth of about 1

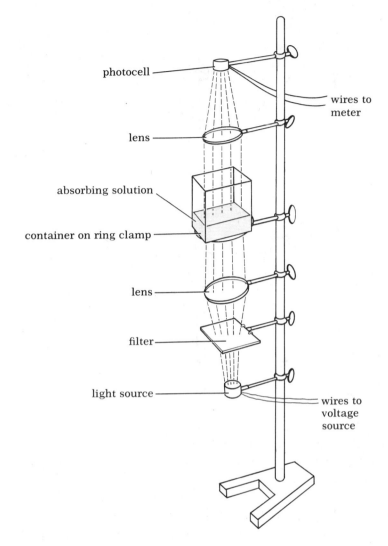

Figure 5-28 The apparatus to study the absorption of light by a solution.

cm. and read the current. The drop should be between 10 and 20 per cent of I_0. If it is not in this range, consult your lab instructor to have the strength of the solution adjusted. Continue increasing the depth and tabulate values of I for different depths x. Before these data take on a meaning, the photocell has to be calibrated.

To calibrate the cell, remove the absorbing solution, the container, and the lenses as shown in Figure 5-29. Adjust the position of the lamp to give the same photocell current I_0 as you had with no absorbing solution. The light intensity on the cell is then E_0. Measure the distance, d_0, between the filament and the photocell. Move the lamp farther away and take readings of photocell current, I, and distance, d. Calculate the intensity E compared with E_0 using the inverse square law. Continue this until you can make a graph of E/E_0 against the current I, which will be your calibration curve. Using this curve and the data concerning the photocell current for different absorber thicknesses, make a table of E/E_0 against the thickness of absorbing solution, x, and make a graph of these data.

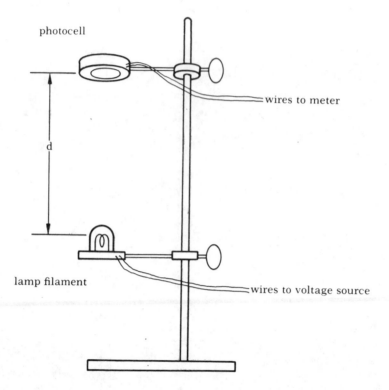

Figure 5-29 The arrangement of the apparatus to calibrate the photocell.

You are not working with ordinary units of light intensity, and to simplify the work call E_0 one unit, so your graph is of E against x.

Use either your data or your graph to show how $\Delta E/\Delta x$ is related to E and if it fits the situation that $\Delta E/\Delta x = -KE$. (The value of ΔE is really negative.) Then proceed to find the relation between E and x and show where the absorption coefficient enters this equation.

Apparatus

Photocell and microammeter
Small stand with right angle clamps
2 lenses—5 cm. focal length with a light shield on at least one of them
Container for absorbing solution—at least 5 cm. deep
Absorbing solution ($CuSO_4$—about 20 grams per liter)
Filter—of a color to pass the light absorbed by the solution
Lamp with a concentrated filament
Steel scale

D. THE DISCHARGE OF A CONDENSER

When a condenser is charged and then a resistance is placed across it, the charge gradually leaks through the resistance and the condenser discharges. But in what way does the charge or voltage (they are directly related by $Q = CV$, where C is called the capacity of the condenser) decrease with time?

The condenser or capacitor is described in Experiment 5-9.

You will use the circuit shown in Figure 5-30. When the switch **S** is closed, the condenser charges and the voltage to which it charges is read on the voltmeter. When the switch is opened, the charge on the condenser flows through the meter. The voltmeter has a high resistance, so in this case it acts as the resistance and also indicates the voltage remaining on the condenser. Read the appropriate part of Chapter Ten on voltmeters to see why. You will also see that to find the resistance of the meter, you must read on it the sensitivity in ohms per volt and multiply by the full scale reading on the range that you are using.

So charge the condenser, read the initial voltage V_0, open the switch, and read V every five seconds until the condenser becomes almost discharged. Try it several times.

Show that $\Delta V/\Delta t = -KV$, and then find the relation between V and t. All of your values of ΔV will be negative.

This situation lends itself to theoretical analysis to find the significance of K. With a voltage V on the condenser, a current i given by $i = V/R$ will flow through the resistance. This results in a reduction of the charge from the condenser by an amount, ΔQ, in a time, Δt. Current, i, is the rate of flow of charge $-\Delta Q/\Delta t$, and for a condenser the charge and voltage are related by $Q = CV$. C is the capacity in coulombs per volt, or farads. This latter relation shows also that a charge, ΔQ, is related to the voltage change, ΔV, by $\Delta Q = C\Delta V$. You have here enough information to show that

$$\frac{\Delta V}{\Delta T} = -\frac{1}{RC} V$$

so $K = 1/RC$ in the relation $\Delta V/\Delta t = -KV$. RC is related to the charge rate and is called the time constant of the circuit.

Put your final equation for V in terms of t in a form involving K and also with K replaced by $1/RC$. Compare your value of $1/K$ with the product RC, where R is in ohms and C in farads or R in millions of ohms and C in microfarads. Also compare the numerical value of the product RC (with R in millions of ohms, megohms, and C in microfarads) with the time in seconds required for the voltage on the capacitor to drop to $1/2.718$ of its initial value. Note also that the product

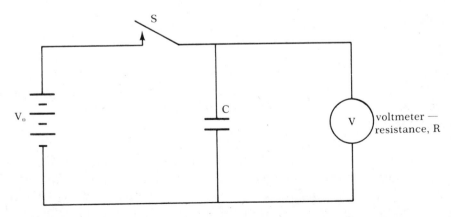

Figure 5-30 The wiring diagram to study the discharge of a capacitor.

of megohms (10^6 ohms) times microfarads (10^{-6} farads) is the same as the product of ohms times farads.

Apparatus

D.C. power supply—100–300 volts

D.C. voltmeter—100–300 volts, at least 20,000 ohms per volt (multimeter)

Single pole single throw switch

Capacitor—3–10 μF, 600 volt rating (choose a value of C in microfarads such that the product CR, where R is the meter resistance in megohms, is in the range of 15 to 25)

Stopwatch

E. GAMMA RAY ABSORPTION

The basic ideas of absorption of radiation are described in the third part of Experiment 5-11 under the heading of The Absorption of Light. In this case, however, you will be measuring the absorption of the invisible, but very penetrating, gamma radiation given off by a radioactive material such as radium or cobalt 60. Your detector will be either a Geiger counter or a scintillation counter, so read the appropriate part of Chapter Ten so that you will see how these instruments function.

With a radioactive source near the counter, familiarize yourself with the counting and reset switches and with the reading of the lights to find the counts in a measured time interval. Your radioactive material gives beta rays as well as gamma rays. In this project you are interested in only gamma ray absorption. A piece of aluminum a quarter inch thick placed directly over the source will eliminate the beta rays sufficiently so that you will count only gamma rays.

To perform the experiment, use the counter in the stand provided as shown in Figure 5-31A and have the source at the bottom, placing the absorbers on a tray near the counter, or use the arrangement shown in Figure 5-31B, with the counting tube held by a burette clamp, the lead absorbers placed on a ring clamp just below it, and the source placed on the desk top. The source will be mounted on an aluminum holder, and it is not dangerous as long as you do not handle it for an extended period. In fact, do not touch the radioactive material itself.

Make several determinations of the counting rate (use several one minute counts) with the source removed to several meters away. You will find that there are always counts. These are due to natural radiation in your surroundings and to cosmic rays. Collectively, they are called the backgound counting rate. Record these counts and find an average background rate, B, which has to be subtracted from all subsequent readings.

Then put the source in position with the quarter inch of aluminum over it. Make several one minute determinations of this counting rate for the radiation intensity E_0. Repeat with lead absorbers, increasing the absorber thickness by a quarter of an inch at a time. Tabulate the data under five headings labeled: lead absorber thickness x, counts C, in time t, rate C/t, and $C/t - B = E$ (the rate minus the background is equal to E).

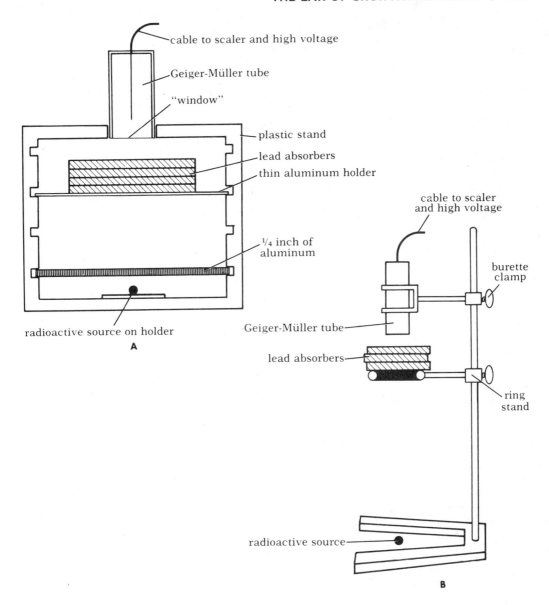

Figure 5-31 Arrangements of the counter tube and absorbers to study gamma ray absorption in lead. In **A** the arrangement uses a stand which may be in a lead box or "castle" to reduce background counts. In **B** a lab stand is used.

Summarize in another table your average of E for each absorber thickness x from $x = 0$ up to your maximum. See if your data show roughly that $\Delta E/\Delta x = -KE$. Your data will probably be quite erratic but make an estimate of K, and then find the equation relating E to x. Where does K, the linear absorption coefficient, occur in this relation?

Apparatus

Geiger counter or scintillation counter with scaler and timer
Appropriate stands

Source of cobalt 60 or radium (1–10 microcuries)
Aluminum absorber—¼ inch thick
Set of 5 lead absorbers—each ¼ inch thick

REFERENCES

1. Earnshaw, S.: Transactions of the Cambridge Philosophical Society. No. 7 (1842).
2. Dieminger, W.: The electromagnetic environment of the atmosphere and nearer space. *In* Benson, O. O., Jr., and Strughold, H., eds.: Physics and Medicine of the Atmosphere and Space. New York, John Wiley & Sons, 1960, pp. 91–94.

6

NUMERICAL ANALYSIS OF DATA

The data obtained in an experiment will be in the form of numbers and it is possible to analyze the data numerically; that is, without resorting to a graph. However, graphical analysis is a quicker and less laborious method of finding an unknown relation between variables. Furthermore, a graph may show very readily any variation from an expected line. On the other hand, the accuracy of a graph is limited. Careful plotting must be done to give results with a precision of even 1 per cent. The value of numerical analysis lies principally in the unlimited precision of the method. The precision of the results will depend entirely on the precision of the data, and not, as in graphical analysis, on the accuracy with which the graph is plotted.

Another advantage of numerical analysis is that the error in the result can be obtained more readily than it can after graphical analysis.

The simplest analysis is of two quantities which are directly related; that is, the relation is of the form $y = bx$, with x and y being the variables and b a constant. To analyze the data, put this in the form $y/x = b$. If this relation actually holds, the ratio of each set of values of y and x should yield the same constant, b. If a number of measurements of the quantities x and y have been made, a corresponding number of values of b will be obtained. Because no experimental measurements are perfect, the various values of b calculated from the ratio y/x will differ from each other, but if there is no general trend in b with changing values of y and x, the theory is not said to be incorrect. (Note the negative form of this statement.) The variations in the calculated values of b show the accuracy to which the relation has been shown to apply in those experimental conditions. For instance, if the various values of b differ on the average by about 1 per cent from their mean, then the relation is shown to apply to about 1 per cent.

For forms of equations that describe other than the simple direct relations, other combinations of the measured quantities can be used. Some illustrations of such relations follow, where a, b, c, and d are constants.

$$\text{If } y = ax^2, \text{ then } y/x^2 = a$$

$$\text{If } y = bx^3, \text{ then } y/x^3 = b$$

$$\text{If } y = cx^{-2}, \text{ then } yx^2 = c$$

$$\text{In general, if } y = dx^n, \text{ then } y/x^n = d$$

From these examples it is seen that the form of the relation must be suspected for this method to be of value.

Madam Curie used numerical analysis when she was investigating the properties of the radiation from the radium she had just discovered. Madam Curie wrote:

> I made several experiments with radium enclosed in a little glass vessel. The rays emerging from the vessel, after traversing a certain space of air, were received in a condenser, which served to measure their ionizing capacity by the usual electrical method. The distance, d, from the source to the condenser was varied, and the current of saturation, i, obtained in the condenser was measured. The following are the results of one of the series of determinations:

d, cm.	i	$i \times d^2 \times 10^{-3}$
10	127	13
20	38	15
30	17.4	16
40	10.5	17
50	6.9	17
60	4.7	17
70	3.8	19
100	1.65	17

> After a certain distance, the intensity of the radiation varies inversely as the square of the distance from the condenser.[1]

Note the cautious wording of her conclusions. At small distances the data do not support the inverse square relation and her conclusion is no more general than the data allow. If an inverse square law held, then the radiation, i, would be related to the distance, d, by an equation of the form $i = k/d^2$, where k is a constant which depends on the amount of radium and the sensitivity of the detector. From this it is deduced that the product $i \times d^2$ would be constant. That an inverse square law should hold is not a necessity for radiation. She found that for the alpha rays from the radium the intensity dropped to zero at about 6 cm. and beyond that there was nothing. This showed that the alpha radiation was not the same as the gamma radiation, which is what was being measured for the above data. The "condenser" to which she refers is what we would now call an ionization chamber. It is the basic instrument for the measurement of radiation dose.

This method, used by Madam Curie to show the existence of a relation between two quantities, can be used also to find an unknown relation. The problem is, given two columns of data as ordered pairs, what combination of the numbers of each pair will give a constant? This may be found by trial and error, and the main ingredient required is perseverance. Kepler took about 20 years to find that for all the planets, R^3/T^2 is the same. No other combination of R and T yields a constant.

As another example, consider the average speeds of the various planets. These are shown in Table 6-1 for the four inner planets. As the mean orbital radius increases, the speed decreases, but what is the relation between them? Remember Boyle's law—as pressure increases, volume decreases, and the product PV is a constant. The planetary situation is similar, and the product vR is shown in column 4 of Table 6-1. Unfortunately, it is not constant, but the list of numbers is still increasing. Again, try multiplying each number of this increasing set by the decreasing numbers in the v column. The result in column 5 is effectively the same for each planet, so v^2R is constant. This is, in fact, a law describing the speeds of objects in various orbits about the same central object.

TABLE 6-1   The Speeds of the Planets and the Radii of their Orbits.
The combination v^2R is constant and amounts to a law describing one aspect of orbits.

Planet	v Units of 10^4 m./sec.	R Units of 10^{11} m.	vR Units of 10^{15} m.^{2}/sec.	v^2R Units of 10^{19} m.^{3}/sec.^{2}
Mercury	4.79	0.579	2.77	13.3
Venus	3.50	1.081	3.78	13.2
Earth	2.98	1.497	4.46	13.3
Mars	2.41	2.280	5.49	13.2

This law that v^2R is constant for any set of orbits has not been named officially, but it is useful. For example, v^2R is the same for all earth satellites in circular orbits. For a low earth satellite, R is about 4000 miles and v is very close to 5 miles per second. The product is

$$v^2R = 100,000$$

For earth satellites, the constant is close to 100,000, using speeds in mi/sec and orbit radii in miles. It is easy to remember for quick estimates of orbit speeds.

Analysis of this type yields an equation showing how two variables are related, but only if the relation involves multiplication, division or powers. If the relation is $y = a + bx$, another approach must be used to take into account a, when only the y and the x are measured. In this case, a series of readings for x and y will be taken, giving:

$$y_1 = a + bx_1$$
$$y_2 = a + bx_2$$
$$y_3 = a + bx_3, \text{ etc.}$$

Subtract adjacent equations to get

$$y_2 - y_1 = b(x_2 - x_1) \quad \text{or} \quad (y_2 - y_1)/(x_2 - x_1) = b$$

and

$$y_3 - y_2 = b(x_3 - x_2) \quad \text{or} \quad (y_3 - y_2)/(x_3 - x_2) = b, \text{ etc.}$$

Again, a number of values of b will be determined. After an average value of b has been found, a set of values of the constant a can be found from $y_1 - bx_1 = a$, etc. The data will then yield values for the constants a and b, and also the spread of the values will show how precisely each has been determined.

It is also possible to analyze data to find an unknown exponent if the equation relating the quantities is of the form $y = bx^n$. Take the log of both sides to get

$$\log y = \log b + n \log x$$

A set of values of y and x would be described by

$$\log y_1 = \log b + n \log x_1$$
$$\log y_2 = \log b + n \log x_2$$
$$\log y_3 = \log b + n \log x_3, \text{ etc.}$$

Subtract adjacent values to get

$$\log y_2 - \log y_1 = n(\log x_2 - \log x_1)$$
$$\log y_3 - \log y_2 = n(\log x_3 - \log x_2), \text{ etc.}$$

or

$$\log(y_2/y_1) = n \log(x_2/x_1)$$
$$\log(y_3/y_2) = n \log(x_3/x_2), \text{ etc.}$$

from which values of n can be found from

$$n = \log(y_2/y_1)/\log(x_2/x_1)$$
$$n = \log(y_3/y_2)/\log(x_3/x_2), \text{ etc.}$$

Such an analysis not only determines n but also shows whether or not n is constant and how precisely that experiment determines n.

THE METHOD OF LEAST SQUARES

An entirely different approach to numerical analysis of data is to solve for the equation of the line which best fits the experimental data. If the data have been plotted on a type of graph on which the points seem to indicate a straight line, the problem is how to draw the best line. Experimental points do not fall on a perfect line, and however it is drawn, some points will be above it and some below. A "best fit" line can be drawn reasonably well with a transparent straight edge. The line in A of Figure 6-1 is obviously a better fit than the line in B. However, there is an element of uncertainty about just which possible line is the best. One type of "best fit" line is the line for which the sum of the distances from each experimental point to the line is less than it would be for any other line that could be drawn. Another form of best fit line is the line for which the sum of the squares of the distances in the y direction from the data points to the line is a minimum—that is, less than it would be for any other line. This form of best fit line can be calculated quite readily, and the method is referred to as the method of *least squares*. The method of least squares will be described briefly here, but a more detailed analysis of it may be found in texts concerned with the analysis of data.

If the distances from the data points along the y direction to the line are represented by E (E_1 for the first point, E_2 for the second, E_i for the ith point, and n for the total number of points), then the sum of the squares of all the E_i's is to be made a minimum. (See Fig. 6-2.) In other words,

$$\sum_{i=1}^{n} E_i^2 \text{ is a minimum.}$$

Consider a set of data points represented by $(x_1, y_1), (x_2, y_2), (x_3, y_3), \ldots,$ the ith point being (x_i, y_i) and the total number of points being n. Let the equation of the line drawn among the points be $y = a + bx$. The y intercept will be a, and b will be the slope.

The distance from this line to the ith point is $y_i - (a + bx_i)$, so the error between the line and ith point is given by

$$E_i = y_i - a - bx_i$$

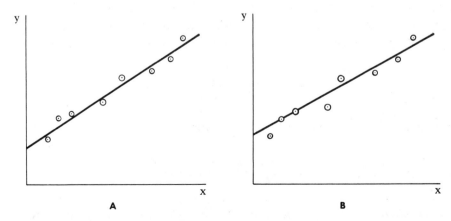

A **B**

Figure 6-1 The "best fit" line. The line in A is a better fit than the line shown in B.

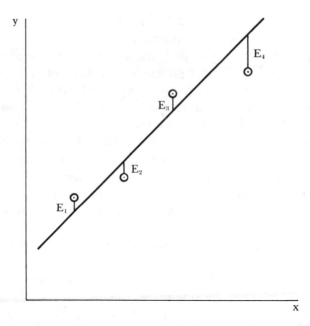

Figure 6-2 The definition of the quantity E_i in the calculation of a "best fit" line. The line is found for which ΣE_i^2 is a minimum.

The square of this is

$$E_i^2 = y_i^2 - 2ay_i - 2bx_iy_i + 2abx_i + a^2 + b^2x_i^2$$

The sum of the squares of all these errors is:

$$S = \sum_{i=1}^{n} E_i^2 = \sum_{i=1}^{n} (y_i^2 - 2ay_i - 2bx_iy_i + 2abx_i + a^2 + b^2x_i^2)$$

This is the quantity that we want to be a minimum. For any line, the slope, a, and the intercept, b, can be chosen independently, so the variations in S will be due to a variation in the choice of a or in a variation in the choice of b. If a is changed by an amount Δa, the variation in S resulting from this is ΔS_a. ΔS_a will depend on the rate of change of S with respect to a (the differential) and the amount of the change, Δa. As a result of a change in a, the change in S is given by

$$\Delta S_a = \frac{ds}{da} \Delta a \text{ (holding } b \text{ constant)}$$

Alternatively, b could be changed by an amount Δb, causing a change in S by

$$\Delta S_b = \frac{ds}{db} \Delta b \text{ (holding } a \text{ constant)}$$

The total change in S is the sum of ΔS_a and ΔS_b. The process of differentiation with respect to one variable at a time was dealt with in Chapter 3; it is called partial differentiation, and the symbol ∂ rather than d is used. Since Δa and Δb will be allowed to approach zero, the calculus notation will be used. With infinitesimal changes in a and in b, the change in S is then given by

$$ds = \frac{\partial s}{\partial a}\, da + \frac{\partial s}{\partial b}\, db$$

Since a and b may be changed independently, ds will be zero only when each of the terms is zero independently. That is,

$$\frac{\partial s}{\partial a}\, da = 0 \text{ and } \frac{\partial s}{\partial b}\, db = 0$$

Neither da nor db is zero, so

$$\frac{\partial s}{\partial a} = 0 \text{ and } \frac{\partial s}{\partial b} = 0$$

Differentiation partially with respect to a (that is, keeping b constant) and then with respect to b (keeping a constant) gives (with all two's canceled)

$$\sum_{i=1}^{n} (a - y_i + bx_i) = 0$$

and

$$\sum_{i=1}^{n} (bx_i^2 - x_iy_i + ax_1) = 0$$

Summing each of the terms in the parentheses gives

$$na - \sum_{i=1}^{n} y_i = b \sum_{i=1}^{n} x_i = 0$$

and

$$b \sum x_i^2 - \sum x_iy_i + a \sum x_i = 0$$

Representing mean or average values by a bar over the quantity allows the above equations to be written more simply. Thus, put

$$\frac{1}{n} \sum_{i=1}^{n} y_i = \overline{y} \qquad \frac{1}{n} \sum_{i=1}^{n} x_i = \overline{x}$$

The foregoing equations can then be written in the following way after dividing through by n:

$$a - \bar{y} - b\bar{x} = 0$$

and

$$b\bar{x}^2 - \overline{xy} - a\bar{x} = 0$$

The slope of the straight line, on the basis of this least squares method, is found by solving the above equations for b and is

$$b = \frac{\overline{xy} - \bar{x}\bar{y}}{\overline{x^2} - \bar{x}^2}$$

The y intercept, based on the same method of analysis, is the solution for a and is

$$a = \bar{y} - b\bar{x}$$

To analyze experimental data in this manner a table similar to Table 6-2 could be used. The values of x and y are the experimental values.

TABLE 6-2 Least Squares Method of Evaluating Data Related to Pressure and Temperature of a Gas at Constant Mass and Volume. These data were taken from a student's notebook.

Temperature in °C. x	Pressure in cm. of Hg y	x^2	xy
0.0	61.7	0	0
14.0	64.9	196	909
20.0	67.0	400	1340
30.5	68.8	930	2098
40.0	71.0	1600	2840
50.0	73.0	2500	3650
60.0	75.4	3600	4524
70.0	77.3	4900	5411
80.0	79.6	6400	6368
90.0	81.9	8100	7371
96.5	83.3	9312	8039
551.0	803.9	37,938	42,550

The values of x and y in Table 6-2 are data obtained from a student's lab notebook and refer to the temperature and pressure of a mass of gas kept at a constant volume. The purpose was to find the temperature at which the pressure would be zero. When the pressure, y, was plotted against temperature, x, a straight line was seen to result. The extrapolation of this line to zero pressure ($y =$ 0) involves first finding the best fit slope, a, and x intercept, b. The equations for these are given above, and the calculation may be carried through with the aid of a table such as 6-2. Each value of x must be squared in finding $\overline{x^2}$, and the products xy must be found in order to obtain $\overline{xy}$. The four columns of Table 6-2 then allow finding $\overline{x}$, $\overline{y}$, $\overline{xy}$, and $\overline{x^2}$. These are substituted in the formulae for a and b. The results are

$$
\begin{aligned}
\overline{x} &= 551.0/11 &= 50.091 \\
\overline{y} &= 803.9/11 &= 73.082 \\
\overline{x^2} &= 37.938/11 &= 3448.9 \\
(\overline{x})^2 &= 50.09^2 &= 2509.1 \\
\overline{xy} &= 42{,}550/11 &= 3868.1 \\
\overline{x}\,\overline{y} &= 50.9 \times 73.08 &= 3660.7
\end{aligned}
$$

slope, $b = (\overline{xy} - \overline{x}\,\overline{y})/(\overline{x^2} - \overline{x}^2) = 207.4/939.8 = 0.221$

y-intercept, $a = \overline{y} - b\overline{x} = 62.0$

x-intercept $= -62.0/0.221 = -281$

The equation of the best fit line is, in terms of pressure P in cm. of Hg and temperature T in $^{\circ}$C., $P = 62.0 + 0.221\ T$.

In a direct relation, the quantities y and x would be the measured values. If a power relation were being investigated, the quantities y and x would be logarithms of the measured values and b would be the exponent. If the relation were exponential, the y values would be the logarithms of one of the measured quantities and x would be the other quantity.

The illustration of the calculation of a best fit line was carried through in detail, and it obviously involves a large amount of calculation. Time can be spent in a better way than just manipulating numbers, and whenever possible a calculator or a computer should be used in the finding of a best fit line. These devices are very useful to save us time for other things, but we must know just what process is being carried out and just what type of best fit line is being found!

CONSERVATION LAWS

The testing to find the validity of conservation laws is another form of experiment analyzed by what would be called numerical analysis. This type of experiment involves the determination of two or more quantities which, if the conservation law holds, will be the same. However, the quantities being compared are obtained from measurements, so they are not exact. Consequently, it is to be expected that the numbers will not be exactly the same. Each of the numbers will have an error but two numbers will not have been shown to be different if, *with their experimental errors*, they overlap. The conservation law will not have been shown to be exact but to be valid only within the experimental error. If the two quantities are not the same within their experimental errors, then the conservation law being tested does not hold in that situation (see p. 23).

Problems

1. According to Kepler's third law, the ratio of the cubes of the mean radii (semimajor axes) of the orbits of the planets to the squares of their periods is the same for each. Below are listed data pertaining to some of the satellites of the earth. Does Kepler's third law describe this situation too? Use a numerical method of analysis to judge this. The reason for any significant discrepancy should be investigated.

Satellite	Mean Radius of Orbit in Miles (Semimajor Axis), R	Period in Minutes, T
Sputnik I	4320	96.2
Explorer I	4711	109.6
Vanguard III	5284	130.
Vanguard I	5388	133.9
Explorer VI	17,176	762.
Moon	238,860	39,343.

2. Take your own data from one of the experiments of Chapter 5 and apply the least squares analysis. Compare the result with that which you obtained by graphical analysis.

3. Use the data from either the experiment on light from a linear source or the one on the field near a dipole to show numerically that the inverse square law does *not* apply in those situations. Then show a combination of numbers that demonstrates the actual relation. This is similar to Madam Curie's presentation.

Resolving Power of a Telescope

Resolving power refers to the ability to see detail. In telescopes it concerns the ability to see detail on a planet, to see individual stars in a distant galaxy, to see characteristics of a quasar, etc. Resolving power is important not only with optical telescopes but also with radio telescopes, spectroscopes, microscopes and many other pieces of equipment. This project is an introduction to resolving power, and it will also show what affects resolving power. The project is in several parts, and to give the whole thing more meaning, each part should be brought to a conclusion before proceeding with the next.

(A) The Phenomenon

The first step is to observe and describe the phenomenon. In this case look at some detailed objects, or at Figure 6-3 from 5 to 10 meters, using a small telescope, and vary the size of the aperture of the telescope using the diaphragm placed just in front of it. At certain aperture sizes the detail disappears. Look at different objects to see how the size of the detail and that of the aperture are related. Describe this in words in your laboratory notebook.

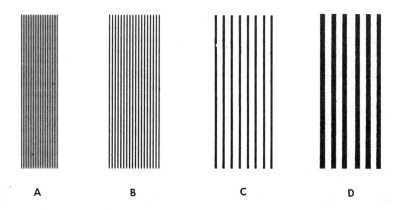

A   B   C   D

Figure 6-3 A resolution chart for obtaining the resolving power of a telescope and of the eye.

(B) Measurements

To study the phenomenon further, the quantities that affect it must be measured and the relation between them found. It is easy to measure aperture size. A caliper or a comparator (lens and scale) could be used, but the precision must be to 0.1 mm. The measurement of resolving power is perhaps new.

The resolving power of a telescope is defined as the angular separation of two objects that can just be seen to be separate. In astronomy it is usually expressed in minutes or seconds of angle. Your final result will be expressed in minutes and seconds, but there are some advantages to doing the initial work in radians.

In this project the object consists of a series of parallel lines such as in Figure 6-3. The angular separation of concern is the angle between two lines when they are viewed from the position of the telescope. If the separation is s and the distance is D, the angle θ in radians is s/D. This is described in more detail in Chapter 10 under Measurement of Angles. If Figure 6-3 is used as the object, it may be placed at 4 to 10 meters. The actual distance should be that for which the smallest aperture available can "blur out" the most widely spaced lines. The angular separations from the viewing positions are found by measuring s and D. Before proceeding further, tabulate the values of s and D in meters and θ in radians.

The next step is to view the various parts of the object and to find the aperture size that will just allow resolution. You will obtain two columns of numbers, one of resolving power, R.P., and one of aperture size, a (use mm).

(C) The Relation Between Resolving Power and Aperture

This relation could be found graphically or numerically. Try this one numerically, as described at the beginning of Chapter 6. You have two columns of numbers, R.P. and a, corresponding to y and x of the text. Find the relation by trying various combinations like R.P./a, etc., as suggested until a combination that gives a reasonably constant answer is found. In a case such as this it is good to tabulate the combinations that were found *not* to work as well as the one that does give a constant value.

You can then write an equation relating resolving power and aperture size. Use the average value of the constant in this equation. For consistency among all who are doing this project, use aperture size in mm. The equation will be in terms of angular resolution in radians, but express it also in terms of angular resolution in seconds of angle. That is, you will have two equations, one in radians and one in seconds. One second of angle is 4.85×10^{-6} radians.

(D) The Role of the Wavelength in Resolution

The experiment is being done with white light for which the wavelength, λ, is in the range from 0.4×10^{-3} mm to 0.7×10^{-3} mm. The mean is about 0.55×10^{-3} mm. Compare this wavelength with the constant in the equation relating R.P. in radians to aperture size in mm., and write your experimental equation for R.P. in terms of wavelength and aperture.

(E) The Resolving Power of Various Instruments

i. Calculate from the equation found in (C) the possible angular resolution in seconds of the 200-inch aperture optical telescope on Mount Palomar.

ii. Find the angular resolution obtained with the 250-foot diameter radio telescope at Jodrell Bank in England when it is used with waves of a length of 21 cm. This wavelength is emitted by a neutral hydrogen atom in space and is commonly measured.

iii. Find the angular resolution obtained at 21-cm. wavelength with the 3300-foot diameter fixed-position radio telescope in Puerto Rico.

(F) The Resolving Power of a Human Eye

An eye also has a limit of resolution, and this can be measured in the same way as was done with the telescope. Move back and forth from the object used to find the telescope resolution. Measure the distance, D, at which a set of lines can just be resolved, and then the angle of resolution, $\theta = s/D$. With the help of another student, somehow make an estimate of the size of the pupil of your eye, the aperture, and determine whether your observed resolution is reasonably close to the results given by the equations found in (C).

The final report should include:

(a) A short written description of the phenomenon.

(b) The columns of data of R.P. and a and the column showing the combination that yielded a reasonably constant value.

(c) The equations relating R.P. and aperture using radians and seconds. (As a rule of thumb, astronomers often use that for an optical telescope. R.P. in seconds is given by 5 divided by aperture size in inches.)

(d) The equation you found relating resolving power, R.P., wavelength, and aperture, a.

(e) The expected resolution in seconds or minutes of a 200-inch optical telescope and of radio telescopes of apertures of 250 feet and 3300 feet for 21-cm. waves.

(f) The resolution of your own eye in minutes of angle and a comparison with the result expected on the basis of the formula you found for a telescope. Visual acuity is defined as the inverse of the resolution in minutes.

This phenomenon of resolving power can also be analyzed theoretically, and you can look forward to the time when you will have a class that includes physical optics, with which an analysis is made. Meanwhile, you have gained an understanding of the limitation of telescopes and the need for large ones.

Apparatus

Small telescopes—6 power, 1-inch aperture (available as "finder" telescopes)
Series of objects—screens or lines such as in Figure 6–3
Series of apertures from 1 mm to 10 mm for telescope, or an iris diaphragm
Meter stick
Comparator or caliper accurate to 0.1 mm.
Stands for telescope and aperture
Lamp to illuminate object

6-2

Inelastic Collisions and Muzzle Velocity

PERFECTLY INELASTIC COLLISIONS

Two quantities associated with a moving mass are momentum (mv) and kinetic energy ($mv^2/2$). These two quantities are associated with the conservation laws known as conservation of momentum and conservation of energy. In this project you will see whether or not either or both of these conservation laws apply in the situation of two objects colliding and then coupling together so that they move off as one after the collision. This type of collision is referred to as being perfectly inelastic. The results will then be used to determine the velocity of a bullet from a rifle.

The apparatus consists of a length of track on which are put small "cars" which will couple together upon collision. To one car is attached an arm which rides just above a waxed tape on a metal bar. A sparking device is used to send sparks 60 times a second from the car through the waxed tape to the metal bar and thereby record the position of the car. The masses of the cars are determined with a balance. The marker must be attached to the car that is initially set in motion. For best results the car initially put into motion should be the lighter of the two.

To perform the experiment, place one car in the center of the track and, with the sparking device operating, push the other one toward it. Remove the tape and measure the distance between successive dots or pairs of dots. Because of friction the velocities continually decrease, but there is a sudden change in velocity at the time of collision. To find the velocities just before and just after the collision, plot a graph of velocity against either time or spark number. From this graph determine the velocities immediately before and after the collision. Then calculate the total momentum just before and also just after the collision. Calculate also the kinetic energy just before and just after.

Refer back to the purpose of the experiment and if the results are inconclusive, repeat the experiment. It is usually faster to obtain a new set of data than to worry over one set.

Apparatus

2 cars arranged to stick upon colliding
Track (could be an "air track" or toy railroad type)
Sparking apparatus and tape
Meter stick

MUZZLE VELOCITY OF A BULLET

To measure the muzzle velocity of a rifle bullet we may use the conservation laws. The bullet is fired into a block of wood which will then move with a sufficiently low velocity to measure with the sparking device. From the first part of this experiment you already know whether the total kinetic energy or the total momentum of the system will be the same after the collision as before the collision.

The mass of the wooden block is determined with a balance. The mass of the bullet is found by removing the lead from several cartridges and finding the mean mass of the lead. The block is mounted on a trolley and rails so that the sparking apparatus can be used to measure its velocity. Because of friction the velocity will continually decrease, but a graph of velocity against time can be used to find the initial velocity. If an "air track" is used, frictional effects will probably be negligible.

For this experiment a .22-caliber rifle may be used with a block of wood 4 × 4 × 12 inches on a moving cart. The rifle should be clamped securely and a backstop should be provided. A closed plywood box, inside of which is a sloping iron plate and in the bottom of which is sand, stops .22 bullets very nicely. You should take care not to load the rifle until the block has been put at rest and everyone is ready and out of the way. The apparatus is illustrated in Figure 6-4.

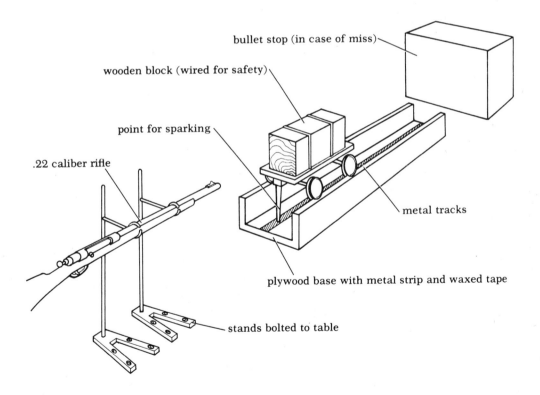

Figure 6-4 The arrangement of the apparatus to measure the muzzle velocity of a rifle bullet.

Apparatus

Block of wood mounted on trolley and rails
.22-caliber rifle and bullets
Sparking apparatus and tape
Balance reading to about 3000 gm.
Balance sensitive to 0.01 gm.

6-3

Elastic Collisions

An elastic collision is one in which there is no change in the total kinetic energy of the objects. Very frequently collisions between atoms or between subatomic particles are perfectly elastic. With objects of ordinary size perfectly elastic collisions do not occur, but with objects such as hard steel ball bearings the collisions are close to being elastic. This purpose of this project is to make some observations on the kinetic energy, momentum, and coefficient of restitution in a collision between ball bearings.

The apparatus consists of a horizontal glass over a blackened box. Just under the glass is a magnet to position the target bearing. Another bearing, the projectile, is fired at the stationary one using a blow tube. The collision is photographed with a Polaroid Land camera under stroboscopic illumination at 24,000 flashes per minute. The experiment consists of taking a picture and then analyzing it with a protractor, steel scale, and microscope or dividers.

It is important that the masses of the bearings be different, or the situation will be a special case and the results may be misleading. It is satisfactory to use a five-sixteenth inch bearing as a target and a quarter inch bearing as a projectile. Weigh each bearing before you start. It will be quite satisfactory to use the distances on the film to calculate velocities, although without a conversion factor to the real velocities, the usual units do not apply.

For each particle before and after the collision find the velocity (v), the momentum (mv), and the kinetic energy ($mv^2/2$).

Find the total kinetic energy before and the total kinetic energy after the collision.

Find the total momentum before and the total momentum after. Remember that momentum is a vector quantity.

Find the velocity of approach, and from a vector diagram find the velocity of separation. The velocity of separation is the difference between the two velocity vectors after the collision. The coefficient of restitution, e, is the ratio of the velocity of separation to the velocity of approach.

Apparatus

Black box with glass top and magnet to hold a target bearing
A blow pipe fastened to the box
Ball bearings—1/4 inch and 5/16 inch
Strobe lamp
Polaroid Land camera
Measuring microscope (or dividers and steel scale)
Protractor
Balance

6-4

Velocity and Frequency of Sound

Does the velocity of sound in air depend on the frequency? Do high notes travel faster or slower than low notes? This experiment is designed to measure the velocity of sound over a wide range of frequencies in order to investigate these questions.

Use is to be made of the relationship among velocity (v), frequency (f), and wavelength (λ), expressed as $v = f\lambda$.

An audio generator with a speaker attached is to be used to produce sounds of different frequencies. The wavelengths are to be measured using a resonance tube.

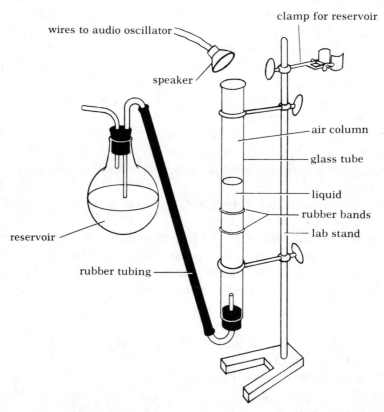

Figure 6-5 The resonance tube apparatus to measure the wavelength and then the velocity of sound at various frequencies. The length of the air column is varied by raising or lowering the reservoir.

One form of resonance tube consists of a large diameter glass tube to which is attached a reservoir as shown in Figure 6–5. With the reservoir held above the top of the tube, the tube is filled with a liquid such as ethylene glycol (permanent type antifreeze) or even water. As the reservoir is lowered, water drains into it from the tube and an air column is produced above the liquid.

To perform the experiment hold your ear and also the earphone near the top of the air column and, starting with the liquid near the top, lower the liquid level fairly rapidly. A succession of resonances will be observed. Have a second observer mark the level of the liquid at the points of resonance. Repeat as the liquid is raised. Repeat several times until the highest and lowest resonance levels have been determined with confidence. Find the average distance between the levels which will be some multiple of half a wavelength, as shown in Figure 6–6. The resonance positions between the highest and the lowest may be determined only approximately in order to find the number of half wavelengths.

Make the first determination at about 500 cycles per second and other determinations at about 100 cycle intervals as far above and below this as possible.

Calculate the wavelength for each frequency and then the product $f\lambda$, which is the velocity at that frequency. Tabulate all these data and, with due consideration of errors, comment on the original questions.

The room temperature should be recorded because the velocity of sound does vary with temperature.

An alternative method to measure wavelengths is to use a speaker, a microphone, and an oscilloscope in addition to the audio oscillator. A second permanent magnet speaker is excellent as a microphone. The speaker and

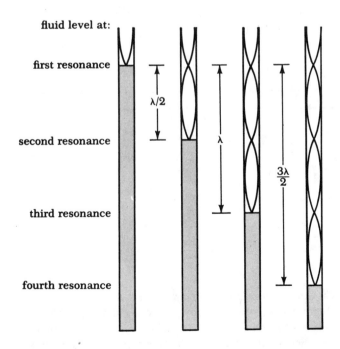

Figure 6–6 Modes of vibration in open-ended air columns.

microphone are set up as shown in Figure 6–7, so that the distance between them can be varied from almost touching to about 2 meters. The output of the oscillator is put into the speaker and also onto the horizontal plates of the oscilloscope. The microphone is put onto the vertical oscilloscope plates. The pattern on the oscilloscope screen will depend on the phase difference between the signals from the microphone and speaker. If the microphone and speaker were together, the waves would be expected to be in phase and the oscilloscope would show a sloping line. As they are separated, the time delay for the sound to reach the microphone causes the vibration detected by it to be out of phase with that fed to the speaker, and the oscilloscope pattern would become elliptical. When the separation is one wavelength, the waves would again be in phase and the pattern would appear again as a line sloping in the same direction. This pattern would repeat at 2, 3, or n wavelengths separation where n is an integer. These patterns are shown at the bottom of Figure 6–7. A measurement of the distance that the microphone is moved between two positions at which the line is obtained on the oscilloscope gives an excellent measure of the wavelengths.

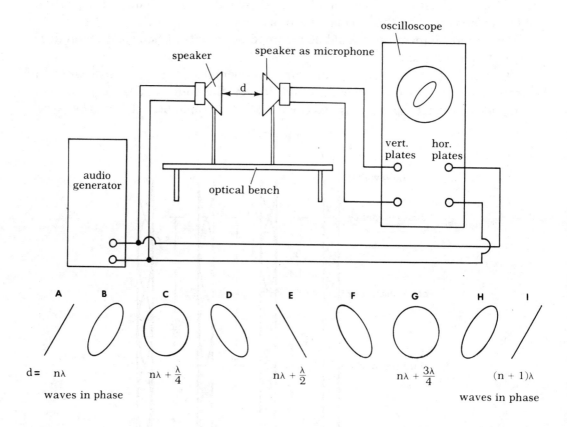

Figure 6–7 The use of an audio oscillator, speaker, microphone, and oscilloscope to measure the wavelength and then the velocity of sound. The various patterns shown along the bottom will be seen on the oscilloscope depending on the separation of the speaker and microphone. The wavelength is the distance between two successive positions at which the pattern is seen as it appears in **A** or **I**.

Apparatus

First method
    Audio oscillator
    Earphone
    Resonance tube apparatus
    Meter stick
Second method
    Audio oscillator
    2 small permanent magnet speakers, one used as a speaker and the other as
        a microphone
    Optical bench
    Oscilloscope

REFERENCES

1. Curie, M.: Radioactive Substances. New York, The Wisdom Library, 1961, p. 48. This is Marie Curie's thesis and one of the most fascinating scientific documents. It is very readable and full of surprises about the thoroughness with which she investigated radioactivity.

7

EXPERIMENTS SUGGESTED BY THEORY

In addition to describing phenomena already observed, theories will usually predict something new to be discovered. Some scientists say that a theory is of no value unless it does predict something new. The theory may then stand or fall depending on whether the predicted phenomenon is as that theory suggested or not. Then another prediction will lead to another experiment, and again the theory is on the judgment stand.

A striking example of this occurred after Einstein, in 1916, showed that one unexpected prediction of his theory of relativity was that a light ray would be bent in a gravitational field.[1] Following his theoretical analysis Einstein wrote that "according to this [prediction], a ray of light going past the sun undergoes a deflection of 1.7″; and a ray going past the planet Jupiter a deflection of about .02″." The deflection of 1.7 seconds is large enough to be measurable for starlight passing the sun if the stars could be seen past the edge of the sun. The predicted deflection of starlight past the edge of Jupiter is less than can at present be measured.

Fortunately, the stars can be seen near the sun during a total eclipse. The first opportunity to test the theory occurred in 1919 when an eclipse was visible in Africa. Another opportunity came in 1922 when an expedition to western Australia obtained some excellent data. Their results were reported in the Publications of the Astronomical Society of the Pacific in 1923 and are now more easily obtained in *Source Book in Astronomy, 1900–1950.*[2] The results of the analysis showed a mean deflection of 1.72 seconds ± 0.11 second whereas a more precisely calculated value according to Einstein was 1.745 seconds. Clearly the experimental value *had not been shown to differ* from the predicted value.

The theory of relativity had stood up well in its first major test. Even today many experiments are being devised to find the limit of applicability of the relativity theory. Each prediction made by it—that mass increases with velocity, that time depends on velocity, that absolute velocity cannot be determined, that light emitted by atoms is somewhat redder if the atom is in a gravitational field than if it is not, and many others—has been observed experimentally, or experiments are being devised for its observation.

As we progress in our physics training, we frequently encounter theoretically predicted phenomena which we should question. As students we can't stop to experiment on every point, but periodically experimental checks must be made so we will see that the theory is, in fact, acceptable. But even then we must realize that the theory under test may not be the only theory that can explain that particular phenomenon. An experiment cannot verify a theory: the only conclusive thing it can do is show that the theory is not satisfactory. This is the reason that there are often rival theories explaining the same thing. As research progresses one of the theories may be found to be inadequate. It may then be discarded or it may be kept for use inside its own limits. Particularly if it is simpler than its rival, it will be kept for limited use. Often one critical experiment is devised to choose between two such theories.

The refraction of light rays as they go from air into a medium such as water or glass can be explained on the basis of the light waves slowing down as they enter the medium (if light is a wave motion), or it can be explained by the theory of the light corpuscles being attracted to the medium as they approach near to it so that the component of velocity perpendicular to the surface is increased. In the case of the wave theory, the speed of light would be *less* in the medium than in air. In the case of the particle theory, the speed in the medium would be *greater* than that in air. The crucial experiment would be to measure the speed of light in air and in a medium. This particular crucial experiment was performed in France by Foucault in 1850. He measured the speed of light in air and in water using his famous rotating mirror method. The result showed that light traveled more slowly in water than in air. The experiment was decisive.

The form of experiment that will test the validity of a theory is not always (or usually) obvious. The theory will have to be worked at, the implication studied, and eventually, often suddenly, an idea will come to mind for an experiment that will show just what is desired. Often the simplicity will be startling and once the idea is there, it is a wonder that it wasn't obvious from the beginning. That is usually the way with the best of ideas. An example of the direct and indirect approach follows.

The resistance force on an object moving in a medium is expected to be of the form

$$R = kv^n$$

The resistance force is R, k is a constant, v is the speed, and n is expected to be 1 for streamline flow and 2 for turbulent flow. The obvious way to investigate n for a particular situation is to measure R as a function of v and then determine n by either graphical or numerical analysis. This direct approach is not the only way to measure the exponent n. It can be shown theoretically that the terminal velocity of an object falling in a medium is proportional to the nth root of the weight in that medium. So the power n can be determined from the measurement of the terminal velocities of objects of different weights. In order to keep the quantity k constant, the objects must all be of the same size and shape.

The analysis is as follows, with the forces illustrated in Figure 7–1. G is the gravitational force, B is the buoyant force, and R is given by kv^n. At the terminal velocity v_t, $R = G - B$. $G - B$ is the weight, W, of the object in the medium. Then,

$$kv_t{}^n = W$$

so

$$v_t{}^n = \frac{W}{k}$$

or

$$v_t = \left(\frac{W}{k}\right)^{\frac{1}{n}}$$

After you find v_t and $(G - B)$ for several objects of the same size and shape but different weight, the quantity $1/n$ and then n itself can be determined using graphical analysis.

It is in the manipulation of a theory in order to find the best ways to obtain meaningful experimental results that the physicist has a great opportunity to show his ingenuity.

If you are dealing with objects for which the resistance force constant k is the same, and with streamline flow conditions so that $n = 1$, the equation above becomes $v_t = W/k$. In other words, the terminal velocity would be proportional to the weight in the medium. This was previously discussed in Chapter One.

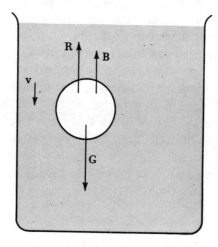

Figure 7–1 A diagram showing the forces acting on a sphere which is falling in a fluid.

7-1

Simple Harmonic Motion

Simple harmonic motion occurs when a mass moves under the influence of a force described by $F = -kx$. The displacement from the equilibrium position is x. The force causing the mass to be held at the displacement x is F, and if the cause of this force is removed, the mass will be made to move toward the equilibrium positions by a similar force, F. The constant k is called the force constant and is given by F/x or by the slope of the graph of F versus x.

Analysis of the motion (see almost any physics text) shows that if a mass, m, is allowed to vibrate under the influence of such a force, it will move with a period, T, which is related to k by

$$k = 4\pi^2 m/T^2$$

or alternately the period is given by

$$T = 2\pi\sqrt{m/k}$$

The force equation $F = -kx$ is a very common form in actual physical situations, for this is obtained with elastic materials and also frequently describes motion in electric or magnetic fields. The period of vibration, being so simply related to the force constant, is then frequently very useful.

The purpose of this project is to investigate a few physical situations by first observing that, for them, the force is a linear function of displacement and then by predicting the period of vibration for a certain mass. After the prediction has been made, the mass will be made to vibrate and the period will be measured. The predicted and measured results are then compared. The reason for any difference beyond a reasonable experimental error should be discussed.

The following situations are to be investigated:

THE SPIRAL SPRING

A spring scale graduated in newtons is provided. Measure the distance to several of the graduations and calculate the force constant k for that spring, or make a graph of F versus x and find k from the slope. Then for a mass that will produce about two-thirds full scale expansion, calculate the expected period.

After this period has been calculated, fasten the mass (for which the calculation was performed) to the hook on the scale. Set the mass vibrating and time 10 vibrations. Repeat the measurement to get at least three consistent determinations; then calculate the average period. Be sure to begin counting at "zero," not "one," the instant that the stop watch is started. Do the predicted and measured periods agree within their experimental errors?

A METAL BAR

A thin steel bar (steel meter scale) is clamped near one end (see Fig. 7-2) so that slotted weights may be rested over it near the free end and can be made to vibrate with a horizontal motion. The spring scale which was used earlier is used here to apply a horizontal force, F, to the end of the bar. The displacement can be measured either with a measuring microscope or by putting another scale on the table as shown in Figure 7-2. The position of the bar can be read by sighting past the vertical edge. Three or four forces should be applied and corresponding displacements should be determined and tabulated.

The force equation for the displacement is shown to be described by an equation near the form of $F = kx$, and the constant k is determined.

Find a mass of such a size that it will vibrate satisfactorily. Using this mass and the measured value of k you can calculate the expected period of vibration by $T = 2\pi\sqrt{m/k}$. Then the mass should be set in vibration and a good measured value of the period should be found.

The calculated and experimental values of the period are then compared. If agreement is not satisfactory, it could be because the mass of the steel bar has been neglected. When m is zero, the period should be zero. So remove the added weights and vibrate the empty bar. You may need a stroboscope to find the period. From this measured period, calculate the effective mass of the bar (m_o) using $T = 2\pi\sqrt{m_o/k}$.

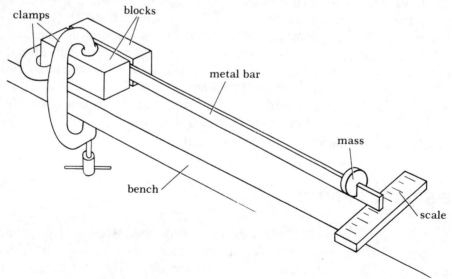

Figure 7-2 The apparatus used to study the vibration of a metal bar.

Add this effective mass of the bar to the mass previously used and recalculate the period. Now how do the theoretical and experimental results compare?

Find the mass of the steel bar and compare the effective mass with the total mass. Does it appear to be a simple fraction?

Apparatus

Spring scale marked in newtons—range of 0-0.3 newtons
Set of weights
Steel meter bar
C clamps and wooden blocks
Stop watch
Balance
Possibly a stroboscope

7-2

The Doppler Effect

The Doppler effect is a change in the observed frequency due to motion of either the source of a wave motion, the observer, or a reflector. Highway patrols utilize the Doppler effect on radar waves reflected off moving vehicles to determine their speed. As satellites are put into orbit, the speed achieved by each rocket stage is determined by the Doppler effect on a transmitted radio wave from the rocket to a ground receiver, and satellite speeds can be measured in the same way. The speeds of stars and galaxies are measured by the Doppler effect on the light from them.

When a source of sound moves with a speed v toward an observer, the observed frequency, f_o, is given by

$$f_o = \frac{V}{V - v} f_s$$

V is the speed of the waves in the medium, and f_s is the frequency of the source. This expression holds for light or radio waves also as long as the speed is small compared to that of light. Even the speed of a satellite, which is about 5 miles per second, is sufficiently small compared to the speed of light, which is 186,000 miles per second. For satellites or automobiles being observed by radar, the change in frequency is a very small per cent of the frequency, but instead of measuring the observed frequency, f_o, the method of beats can be used to measure the change in frequency $f_o - f_s$. It is only by this technique that the method of speed measurement with radio waves is possible. Figure 7-3 is a graph showing how the frequency of a radio wave from a satellite changed as the satellite approached and then receded from an observer. The total change was only about 1000 cycles per second for a transmitted frequency of 20 megacycles per second. In this instance the signal received was beat against an oscillator on the ground, the frequency of which was very close to the frequency being broadcast from the satellite. This diagram resulted from measurements by two radio "hams."[3] As the satellite approached, the frequency received was about 1920 cycles per second above the reference oscillator frequency, then as it receded from the observer the frequency dropped to about 960 cycles over the reference. It is left as an exercise for interested students to calculate the satellite speed from these data.

The object of your experiment is to measure the speed of a sound source by using the Doppler effect. The method employed is based on the same principle as that used to measure satellite speeds, except we will use sound waves instead of radio waves. The speed can be measured with a stop watch also so that the results from the Doppler measurement can be checked.

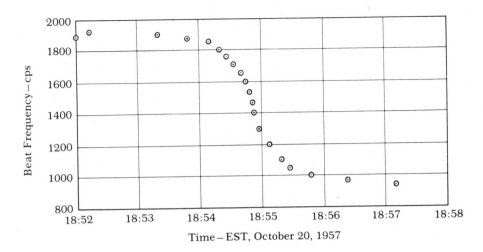

Figure 7-3 A graph of the change in frequency of the signal from a satellite (Sputnik I) as it passes an observer. (From Peterson, A.: General Radio Experimenter. *32*:3 (1975). Used with permission of General Radio Company, West Concord, Massachusetts.)

The apparatus consists of an audio oscillator and two speakers, one of which is on wheels. A motor is adapted to pull the wheeled speaker along the table. The other speaker is held stationary and serves as a reference by emitting the same frequency as the moving source would if it were stationary. When one speaker moves, beats are produced, that is, the intensity seems to rise and fall. This phenomenon of *beating* occurs when two different frequencies are combined. Briefly, it results from the waves being sometimes in phase and sometimes out of phase at the detector (ear, microphone, etc.). If the frequencies differ by one cycle per second, the waves will become out of step, or out of phase, once every second, giving one *beat* per second. If they differ by two cycles per second, they become out of phase twice per second. The beat frequency is the same as the difference between the two frequencies. Beats can be used to measure very small frequency differences. To find more information on the topic of beats, look up that term in the index of a physics text.

In this case the beat frequency is the difference $f_o - f_s$, where f_o is the frequency observed from the moving speaker and f_s is that observed from the stationary speaker. The beats can be counted and timed with a stop watch as the speaker moves along the table. The experiment is best performed by two observers. One counts and times a number of beats as the speaker moves and the other times the speaker as it passes over a marked and measured distance along the table. Several trials should be made at varying speeds, and then at one of the best speeds for counting beats, several determinations should be made. Observers should also switch around. One method of counting the beats is to mount a microphone at the end of the track and display the output on an oscilloscope. The beats appear as a rise and fall of the display pattern. To eliminate room noise, a filter which will pass only a narrow band of frequencies covering the frequency used may be necessary. A block diagram of the apparatus required is shown in Figure 7-4.

The equation given for the Doppler effect can be solved for v. This must be shown in your report. To calculate v, the source frequency (f_s), the velocity of

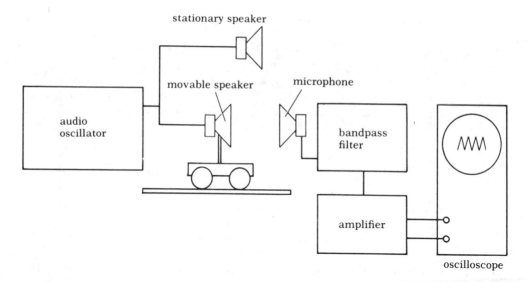

Figure 7-4 The apparatus used to make measurements of the Doppler effect on sound.

sound (V), and the frequency difference ($f_o - f_s$) must be known. The source frequency can be obtained from the oscillator setting. The velocity of sound is found from $V = V_o \sqrt{1 + T/273}$, where V_o is the velocity of sound at 0°C. and T is the centigrade temperature of the air. The frequency difference is the beat frequency.

Calculate the speaker speed by the Doppler method and compare this result to the measured speed.

Apparatus

Audio oscillator
4-inch speakers—one stationary and one on a wheeled cart
Variable speed motor—adapted to pull the speaker
2 stop watches
Meter scale
Thermometer
Optional
    microphone
    bandpass filter
    amplifier
    oscilloscope

7-3

Speed of Electrons

Outline and Theory

Electrons near the surface of a metal can be ejected into the space around it when the metal is heated. These electrons have shared the thermal motion of the atoms. They are what make up the beam in a TV tube, in an oscilloscope, in an x-ray tube, in a thermionic diode, and in many other instruments. The purpose of this project is to measure the speed of the electrons emitted from a hot metal.

The project introduces some concepts that are also applicable to other areas. The atoms of the metal have transmitted their energy to those electrons, and, hence, the measure of kinetic energy of the electrons gives a measure of the kinetic energy of atoms at whatever temperature the metal might be. If the energies are equal, the speeds correspond as the square roots of the masses. Air molecules are about 53,000 times as heavy as electrons, and their speed is consequently about one-230th of the electron speed for that temperature. You are to calculate the electron speeds and convert them to motion of air molecules at that red-hot-metal temperature. Some comments can then be made about high-speed aircraft and the re-entry problem for space vehicles. In these cases the vehicles move toward the molecules at high speeds, and the effect is almost as if the air were moving around a stationary vehicle, which corresponds to a high temperature.

The apparatus to measure the speed of the electrons contains a small metal cylinder that is heated by a tungsten filament inside it. Around this is a metal cylinder or "plate" to collect the electrons. It is, in reality, a thermionic diode. Ordinarily it is used to restrict the passage of current to only one direction. Appreciable current flows only when the hot cathode is negative and the plate is positive. If the plate is made negative and the hot cathode positive, no current flows through it because no electrons leave the plate. This is not quite true if very small currents are considered. With the plate and cathode at the same voltage, electrons ejected from the hot cathode will reach the plate. If the plate is made very slightly negative, the ejected electrons will be repelled and the slow ones will be turned back. With the plate more negative, more electrons will be repelled back to the cathode. By using successively more negative voltages on the plate, more electrons will be stopped. Each voltage will stop electrons up to a given speed, and the current indicates the number of electrons moving at speeds greater than that given speed.

To relate the electron speed and stopping voltage, use is made of the basic definition that 1 joule is required to move one coulomb of charge across 1 volt, and the energy in joules required to move a quantity of charge q across a voltage V is given by

$$\text{Energy} = qV$$

If the particle being moved is an electron, the amount of charge is the electronic charge e, and

$$\text{Energy} = eV$$

If the electron is moving with a speed v, its energy of motion is

$$\text{Kinetic energy} = \tfrac{1}{2}mv^2 \quad (m \text{ is the electronic mass})$$

This energy will carry it across a maximum voltage given by

$$\tfrac{1}{2}mv^2 = eV$$

Alternatively, if the electron is just stopped by a voltage V, its initial velocity is found by solving the above equation for v

$$v = \sqrt{\frac{2eV}{m}}$$

e is 1.602×10^{-19} coulombs

m is 9.1×10^{-31} kg

To save some time in this project, Table 7–1 has been calculated on the basis of this equation, showing a series of electron speeds and the voltages required to

TABLE 7–1 The Voltages That Will Stop Electrons up to a Given Speed. The voltage is in volts, and the speed is in units of 10^4 meters/second. 10^4 meters/sec. = 6.21 mi./sec. = 22,400 mi./hr.

Volts, (V,)	Speed (v,) Units of 10^4 m./sec.
0.01	6
0.04	12
0.09	18
0.16	24
0.25	30
0.36	36
0.49	42
0.64	48
0.81	54
1.00	60

stop them. The velocities shown increase by constant increments. The corresponding stopping voltages increase according to a square law. From the table it is seen that a voltage of 0.01 volts stops electrons up to 6×10^4 m/sec, while a voltage of 0.04 volts stops those up to 12×10^4 m/sec. The difference between the currents at 0.04 volts and 0.01 volts shows the number of electrons moving at speeds between 6×10^4 and 12×10^4 m/sec. The difference in the current with voltages of 0.04 and 0.09 volts shows the number of electrons moving at speeds between 12 and 18×10^4 m/sec., and so forth.

Experimental Procedure: Electron Speeds

The diode is provided with a mount and a 6.3-volt transformer, which need only be plugged into the wall outlet to heat the filament and hence the cathode. The connections to the cathode and plate are marked. The schematic wiring diagram is shown in Figure 7–5. The order of wiring may be as follows:

1. Connect the low voltage source across the ends of the resistor that has a variable tap (rheostat).

2. Connect the voltmeter from the positive end of the voltage source to the variable tap, and with the voltage turned on, note how the output may be varied.

3. Connect the positive terminal to the cathode.

4. Connect the variable terminal (negative side of voltmeter) through a microammeter to the plate.

5. Plug in the filament transformer to heat the cathode, and turn the voltage on.

6. Vary the voltage. The current should be maximum at zero voltage and decrease as the voltmeter reading increases. If the current increases as the voltage increases, the plate is probably becoming positive. Reverse the voltage source and try again.

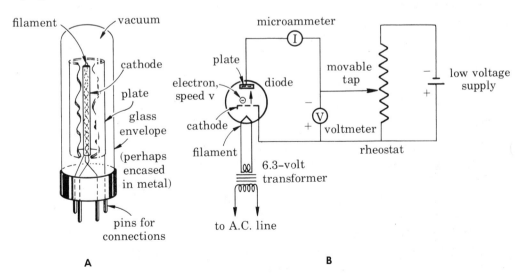

Figure 7–5 **A,** the thermionic diode and **B,** the schematic wiring diagram for finding the speeds of the electrons emitted from the hot cathode.

When the apparatus is working, make a table of current readings at the voltages indicated in Table 7-1. Read to tenths of microamps, and increase the voltage until the microammeter indicates no current, that is, until all the electrons are stopped. Subtract adjacent current readings to obtain an indication of the number of electrons N at each speed interval indicated by Table 7-1.

Make a table of the current difference (or number of electrons N) against speed, using the speed at the center of each interval. Transfer these data to a graph so that you can see the distribution of speeds.

Find the speed at the maximum of the curve.

Find the average speed by multiplying the quantities $N \times v$, adding them, and dividing by the sum of the quantities N.

Find the r.m.s. (root mean square) speed by adding $N \times v^2$, dividing by the sum of N, and taking the square root. This is also designated by $\overline{v^2}$.

The temperature is related to the r.m.s. speed by

$$\tfrac{1}{2}\, m\, \overline{v^2} = kT$$

m is the particle mass, 9.1×10^{-31} kg for electrons.

$\overline{v^2}$ is the square of the r.m.s. speed in m/sec.

k is Boltzman's constant, 1.38×10^{-23} joules/$^\circ$K.

T is the absolute temperature ($^\circ$K).

Calculate the filament temperature.

Calculate the r.m.s. speed of air molecules at that temperature.

Comment on the problem as it concerns high-speed aircraft.

Further work:

The experiment should be repeated using a filament voltage reduced by about 20 percent. The velocity distribution at the lower voltage, and, hence, cathode temperature, is fascinating. It should be plotted on the same graph as that for the full-filament voltage.

Apparatus

Thermionic diode with separate cathode, such as a 6H6 vacuum tube.
Voltmeter with scale divisions at least as small as
    0.1 volts
Microammeter with divisions at least as small as 1 μA
Connecting wires
6.3-volt transformer for filament
Optional
    variable transformer to change filament voltage
    optical pyrometer for actual measurement of cathode temperature if the tube used has a glass envelope.

REFERENCES

1. Einstein, A.: Die Grundlage der Allgemeiner Relativitats Theorie. *In* Lorentz, H. A., Einstein, A., Minkowski, H., and Weyl, H.: The Principle of Relativity. New York, Dover Publications, p. 163.
2. Shapley, H., ed.: Source Book in Astronomy, 1900–1950. Cambridge, Mass., Harvard University Press, 1960, pp. 351–355.
3. Peterson, A.: On the tracking of satellites. General Radio Experiments. *32*:3 (1957).

8

EXPERIMENTS TO DETERMINE CONSTANTS

An experiment designed to measure a constant will differ in some fundamental ways from an experiment designed to find or check a relation between variables. The experiment designed to measure a constant will make use of relations which have already been accepted; then the experiment will be designed to obtain high accuracy under only one or two sets of conditions.

Consider the different ways in which the value of the acceleration due to gravity, g, can be obtained using a simple pendulum. The period of the pendulum, for small amplitudes, is given by $T = 2\pi\sqrt{l/g}$. This relation can be used to find constant g. One method, and perhaps the most obvious, is to solve for g, getting $g = 4\pi^2 l/T^2$. For only one length, l, of the pendulum, the period T can be measured and g can be calculated. The effort of the experiment will be devoted to accurately measuring the quantities l and T.

This numerical method is contrasted to the procedure to be followed in finding that the period is proportional to the square root of l. To find the relation, a large number of values of T over a wide range of lengths l must be measured. The relation between T and l can then be determined by one of the graphical plots described in Chapter Five. The graph can then be used to find g. If the period T has been plotted against $\sqrt{l}$, the graph would be a straight line of slope $2\pi/\sqrt{g}$. Then if the slope is measured, g can be calculated. If $\log T$ had been plotted against $\log l$, the slope of the graph would be one half (the exponent of l), but the y intercept would be $\log 2\pi/\sqrt{g}$. In this case g could be calculated from the intercept. With a graphical method, however, the experimental effort is not used most economically for finding the constant. The slope of a line can be most accurately determined if the points are as far apart as possible, because the error in the slope depends upon both the absolute error in the points and their distance apart. This is illustrated in Figure 8–1, in which the errors of the experimental points are shown by the short vertical lines. The experimental errors are the same

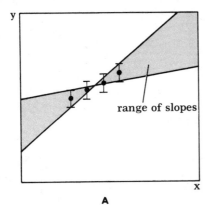 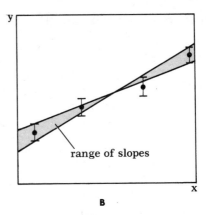

Figure 8-1 These graphs show that only the end points determine the range of possible slopes of a graph and that for a given absolute error the farther apart these points are, the smaller the possible range of slopes.

in both cases, but the slopes of the possible lines that could be drawn lie within a much smaller range in graph **B**, where the outermost points are farther apart, than they do in graph **A** where they are close together. The points toward the middle have little effect on the slope. The effort of the experiment should be devoted to obtaining two sets of data as far apart as possible and as accurately as possible, not in obtaining a large number of in-between points which are necessary to show the linear relation. If two sets of data were obtained, T_1 at a length l_1 and T_2 at a length l_2, the slope, S, of the line between the two points would be

$$S = (T_2 - T_1)/(\sqrt{l_2} - \sqrt{l_1})$$

and

$$S = 2\pi/\sqrt{g}$$

from which

$$g = 4\pi^2 (\sqrt{l_2} - \sqrt{l_1})^2 /(T_2 - T_1)^2$$

For given absolute errors in l and T the error in g can be reduced to a minimum by making the subtracted quantities as large as possible. To help achieve this the smallest possible value of l_1, which is zero, would be used and the corresponding value, T_1, would also be zero. Only one value of l and T remain, so the subscripts can be dropped and the expression for g becomes

$$g = 4\pi^2 \, l/T^2$$

This is identical to the expression used for the numerical analysis. A similar conclusion would be reached if the log graph were analyzed.

Incidentally, a theoretical expression for g has been derived for any latitude θ, or altitude h. This assumes a homogeneous earth and variations from the theoretical value tell about local irregularities. The expression, taken from Childs[1] is

$$g = 9.80616 - 0.025928 \cos 2\theta$$
$$+ 0.000068 \cos^2 2\theta$$
$$- 0.000003\ h.$$

h is the elevation in meters above sea level; g is in m/s^2. When you experimentally measure g, compare your value, with its error, to the value obtained using this expression.

The conclusion is that it is possible to use graphical methods to determine constants. Graphical methods are, in fact, often used in that way, but only if it is necessary to show at the same time that the relation assumed is actually valid for the conditions of the experiment. If the relation can be assumed or has been shown separately to be satisfactory, then the experiment may be devoted to obtaining one or two sets of data and using numerical analysis.

Measuring the same constant in different ways serves to give a more reliable estimate of the constant than if it were obtained in only one way. Furthermore, it then gives an indication of the validity of the assumption used in the different methods.

One of the interesting tales of modern physics concerns the determination of the charge carried by the electron. This charge had been determined by the Millikan oil drop apparatus and also by a method based on x-ray spectra. There was a significant difference in the results obtained by the two methods and it was only after many years of searching that reason for the discrepancy was found. It turned out that the value accepted at the time for the viscosity of air, a quantity which entered the calculations for the oil drop method, was in error. When a newly determined value was used, the values for the electronic charge as calculated by the two methods agreed.

This example shows also how modern physics requires classical physics for its progress. Though not as romantic as outer space experiments, the measurement of physical quantities and constants still is an important part of physics. Such things as density, viscosity, and surface tension may seem ordinary, but there is an exciting side to them also. In the study of behavior of matter at very low temperatures, these quantities are important.

In an introductory laboratory course the way to measure some of these and other quantities at or near room temperature will be learned. The beginning physics student does not work near absolute zero—just as the apprentice mountain climber does not tackle the Matterhorn or Mount Everest. Nevertheless, he is thrilled with the progress he makes on the simpler slopes and peaks that have been climbed many times before.

So some of the experiments described at the end of this chapter will be "how to do it" experiments. From them you will learn how to measure a few of the physical constants or quantities. There have been many excellent and far more comprehensive physics laboratory books to which reference can be made if it is necessary to measure other quantities. The experiments here are illustrative of methods.

It is important that when you have measured a constant you also indicate the precision of that measurement. The calculation of errors has been discussed in Chapter Three.

8-1

Air Thermometer

Procedures to measure low temperatures are based on the properties of gases, and the purpose of this project is to measure the temperature of solid carbon dioxide (dry ice). An ordinary mercury type thermometer cannot be used because mercury freezes at about $-39°C$. The device used will be a bulb of air of constant volume; and as the absolute temperature changes, the pressure will change in direct proportion (Charles' law).

The apparatus to be used is shown in Figure 8-2. The pressure on the gas in the bulb is barometric pressure plus the height of the column, which is shown as h. It is satisfactory to express the pressure in cm. of mercury. For a given mass in a constant volume, Charles' law can be expressed as $T = kP$. T is the absolute temperature, P is the absolute pressure, and k is a constant that depends on the mass and volume of gas. The official conversion between degrees Kelvin, T, and degrees Celsius (often called Centigrade), t, is $t = T - 273.16$.

In this project Charles' law is assumed, and to evaluate the constant k for your apparatus, P must be measured at some known temperature. A convenient

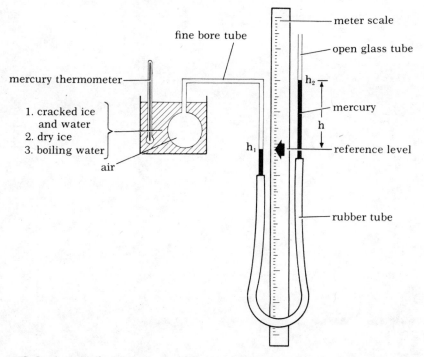

Figure 8-2 An air thermometer which operates at constant volume. The temperature change is determined by the change in pressure.

temperature to use is the freezing point of water, which by definition is 273.16°K. The idea is to measure P with the air bulb immersed in an ice-water mixture and determine k. The bulb can then be used to measure other temperatures simply by measuring the pressure P, which keeps the volume constant, and using $T = kP$.

To measure the pressure with the bulb immersed in cracked ice and water (0°C or 273.16°K), adjust the movable side of the mercury column to keep the level in the tube at the reference mark. (Why?) When it reaches a steady condition, read h of Figure 8–2 and calculate P. You will also have to obtain the barometric pressure from a barometer in the laboratory.

While you are calculating the pressure and the constant k, the bulb can be left in the air to measure the air temperature.

The most important part of the project is to measure the temperature of solid CO_2. Immerse the bulb in a beaker or a Dewar flask of solid CO_2 chunks and measure P and T. You must be alert in doing this, for the air in the bulb contracts quickly, and the mercury may be forced into the bulb, requiring that the apparatus be repaired.

Another suggested measurement (just for curiosity) is that of the temperature of boiling water at the altitude of your lab. Also to illustrate some of the problems of the United States Skylab project in 1973, spray a bit of black paint on one side of the bulb, put it about a foot from a hundred-watt incandescent light bulb, and measure P and T. Put a piece of aluminum foil between the two bulbs to act as a heat shield, and again determine the temperature.

Your report will conclude with a summary of the constant k and a listing of the various temperatures.

You may realize that the bulb expands and contracts, so the volume is not quite constant. If you know the coefficient of expansion of the glass of the bulb, you could calculate the actual volume change. Then using the general gas law in the form that for a constant mass (and number of moles) of gas the quantity PV/T is always the same, the final temperature can be found. Using the subscripts 1 to indicate the values when ice and water is used and 2 for the values at some other temperature,

$$\frac{P_1 V_1}{T_1} = \frac{P_2 V_2}{T_2}$$

The unknown is T_2, *and solving for it*

$$T_2 = \frac{P_2 V_2}{P_1 V_1} \cdot T_1$$

The volume V_2 at the second temperature is different from V_1 by only a small amount; call it ΔV and let

$$V_2 = V_1 + \Delta V$$

or

$$\frac{V_2}{V_1} = \left(1 + \frac{\Delta V}{V_1}\right)$$

Then

$$T_2 = \frac{P_2}{P_1} T_1 \left(1 + \frac{\Delta V}{V_1}\right)$$

In the main part of the project the temperatures were calculated on the basis of only

$$T_2 = \frac{P_2}{P_1} T,$$

but these should be corrected, considering the fractional volume change, $\Delta V/V$. This requires knowledge of the coefficient of expansion of the glass of the bulb (for Pyrex the linear coefficient is about 3×10^{-5} per degree K, and the volume coefficient is three times this.) You should try to show that by increasing the radius of a sphere by a fraction, $\Delta R/R$, the fractional increase in volume is $\Delta V/V = 3\Delta R/R$. The only problem is that to calculate $\Delta V/V$, the second temperature, T_2, must be known. That however, is what you want to find.

The approach to this type of obstacle is to make an intial estimate of T_2 as was done in the first part of the project. Use this to calculate the fractional volume change, $\Delta V/V$ (it may be negative). You then get a better estimate of T_2, and this can be used to get another estimate of $\Delta V/V$. If this is significantly different from the first estimate, use this new value of T_2 to find a new ratio $\Delta V/V$. Continue this until you are content with the answer. Your results will give you even more satisfaction if you have made the correction for the volume change.

Apparatus

> Air thermometer
> Beaker
> Cracked ice (pieces of solid H_2O)
> Dry ice (solid/CO_2)
> Barometer (one per lab)
> Large beakers or Dewar flask
> For optional parts
> > hot plate or Bunsen burner
> > black spray-paint
> > lamp with 100-watt bulb
> > aluminum foil

8-2

The Determination of g by Means of a Simple Pendulum

The object of this experiment is to determine the quantity referred to as the acceleration due to gravity, g. A simple pendulum will be used and the relation $T = 2\pi\sqrt{l/g}$ will be assumed to describe the relation between the period T, the length l, and g. This relation is sufficiently accurate for small oscillations.

The above relation solved for g gives

$$g = 4\pi^2 \ l/T^2$$

The relative error in g is the sum of the relative error in l and two times the relative error in T. π can be used to a sufficient precision so that the error due to its being rounded off is negligible.

The pendulum will consist of a metal sphere on the end of a string suspended from a special clamp.

Make the length l approximately as long as the available stand and measuring scale will allow. If a meter stick is available, use between 95 and 100 cm. Be sure that the string is attached to the clamp in a manner such that the length is the same when it swings in both directions. The construction of a suitable clamp is shown in Figure 8-3. Use your own ingenuity, a meter scale, and a vernier caliper to find the length as measured to the center of the bob.

To measure the period, set the stand vertically and put a line on the table under the bob. When counting swings, count each time the bob goes over this line in a certain direction. Allow the pendulum to make oscillations of only about 5 degrees and be sure to begin counting at "zero" and not "one" when the watch is started. Or else use the rocket launching method of counting 5, 4, 3, 2, 1, 0, 1, 2, . . . and start the watch at "zero." It is a good idea to obtain an estimate of the period by timing just 10 vibrations, and then time 100 vibrations at least twice. If the times do not agree closely, repeat the timing procedure.

List the values of l and T with their estimated errors. Calculate g and the error in this calculated value based on the errors in l and T. How does your measured value compare with the value calculated using the expression given on page 146? If your result, with its error, does not include this value, what may some of the reasons be?

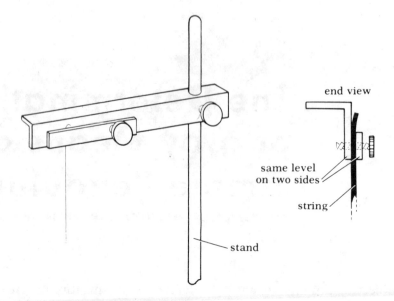

end view

same level
on two sides

string

stand

Figure 8-3 A type of pendulum clamp which assures that the length of the string is the same for both directions of swing.

Apparatus

Large stand
Pendulum clamp
2 yards of fishline
Metal ball—3/4 inch diameter, drilled on a diameter
Steel meter scale
Vernier caliper
Stop watch

The Physical Pendulum to Measure *g*

A physical pendulum is one which has a finite size as contrasted to the simple pendulum, which is ideally a point mass on a weightless inextensible string. The conditions for a simple pendulum are never met very well so a good value of *g* cannot be obtained using it. However, approximations do not enter the analysis for a physical pendulum, so more satisfactory values of *g* can be found. Any shape can be used for a physical pendulum, but it is preferable to use one for which the moment of inertia can be easily obtained.

This is the type of situation in which, as far as the lab work is concerned, a developed relation is used, and one of the quantities in it is found with as much precision as your best effort will allow. In this instance the theoretical development shows that quantity to be the same as the acceleration of a freely falling object. This is the gravitational field strength modified by the rotation of the earth. The importance of the project is that it is an exercise in precision. The measurements of length and of time should be as precise as possible, and be certain to include all corrections that you are aware of.

To achieve the measurement, one of the pendulum shapes described in Figure 8-4 or even others can be used. The distances may be measured with a calibrated steel scale, with a traveling microscope, or even with your lab standard and two measuring microscopes as shown in Figure 8-5. The measuring microscopes are used to compare the measured length with the standard. Consider also the expansion of the scale due to temperature. Strive for an error of ±0.01 per cent or less.

The timing of the oscillations will require special techniqes and one of the methods which follow can be used. The precision of stop watches can be notoriously bad because they are usually run for only short intervals. A mechanical type of clock or watch may have a precision of a minute per day, which is less than 0.1 per cent, but the variation during the day may be more than this. An electric watch may be of greater precision. Whatever timing device is used, it should be (or have been) calibrated against a time standard such as is broadcast by WWV or CHU in North America and GBR or MSF in Great Britain. These time signals are easily available on a short wave set. They have the further advantage of giving "clicks" at one second intervals.

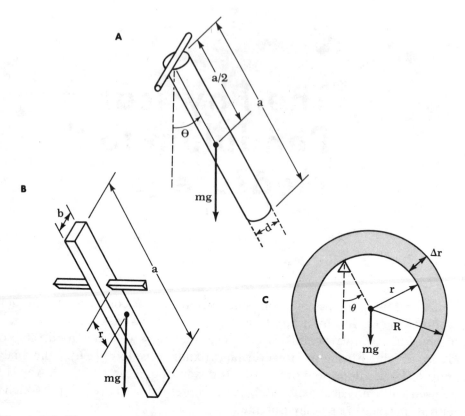

Figure 8-4 Three types of physical pendulums: **A,** A rod pendulum oscillating about one end. **B,** A bar pendulum oscillating about a point at a distance from the center which is equal to the radius of gyration. **C,** A ring or hoop pendulum.

One timing procedure is to first get a fair idea of the period using a calibrated watch and finding the time for 10 swings a few times; then make several measurements of 100 swings. These should agree, especially if you use the "count-down" method, going 5, 4, 3, 2, 1, 0, 1, 2, ... and starting the timing on "zero." It is hard to count more than 100 oscillations, but a trick you can use is to time accurately a whole number of oscillations (about 500) but don't even count them. Use your already determined period to calculate the actual number.

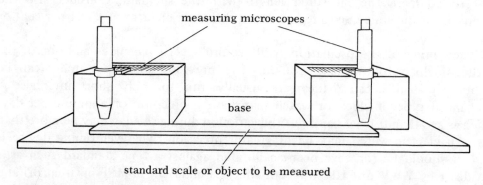

Figure 8-5 How two measuring microscopes are used to compare a 1 meter standard with a length which is close to 1 meter.

If it comes out to 489.1, you know that it was probably 489.0 since you timed a whole number of oscillations. Use this calculated number to get a new value for the period. Repeat this several times and perhaps try in the region of 1000 vibrations also.

There is another timing method that is useful if the period is very close to one second or if it is very close to two seconds: time half periods. The procedure is to first determine if it is less or more than one second. Use a timing device which can give one second beats. This can be a student with an accurate watch or even short wave radio station such as WWV. Tap out the pendulum beats and record the clock times at which the pendulum beats coincide with the clock beats.

For example, if the pendulum is slower than the clock and coincides four times in 19 minutes (1140 seconds), it means that in 1140 seconds there were only 1136 oscillations.

The calculation of g is done as follows for the general case of an extended object supported at a distance h from the center of mass as shown in Figure 8-6. If it is displaced through an angle θ, there is a restoring torque, L, described by $L = mgh \sin \theta$. If θ is small, $\sin \theta$ approaches θ (in radian measure). mg is constant and can be represented by k. Then the torque is given by $L = - k\theta$, which is analogous to the force equation $F = - kx$, which leads to linear simple harmonic motion. In this instance angular simple harmonic motion results and the mass oscillates with a period given by $T = 2\pi\sqrt{I/k}$. I is the moment of inertia about the axis of oscillation given in terms of the moment of inertia about the parallel axis through the center of mass, I_{cm}, by $I = I_{cm} + mh^2$.

Incidentally, work with length in meters and time in seconds to get g in meters/second^2.

The following are analyses of some specific pendulum configurations:

1. A uniform round rod pivoted at one end. (See Fig. 8-4A showing a rod of length a, diameter d, and mass m.) This type of pendulum is simply made and of amazing accuracy. The rod may consist of a welding rod or other rod about 1/8-inch (3 mm) in diameter and of a length short enough to be measured with one measuring microscope, perhaps about 25 cm long. The pivot can be a fine

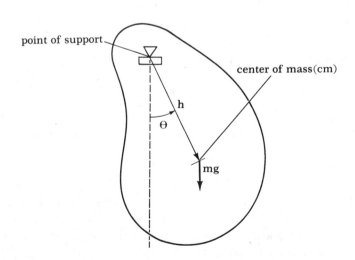

Figure 8-6 A general physical pendulum showing the forces and distances used in the analysis for the period.

needle soldered across one end and supported across a test tube with the bottom removed or across a section of glass rod. A burette clamp is convenient for holding the glass support. Although the top may not be level, when the pendulum is set swinging it will "walk" around the tube until the two ends of the needle are in a horizontal position.

$$I_{cm} = m\,(a^2/12 + d^2/16)$$

$$I = m(a^2/3 + d^2/16)$$

$$L = mg\,a/2 \cdot \theta \text{ for small } \theta$$

$$k = mg\,a/2$$

$$T = 2\pi\sqrt{\dfrac{(a^2/3 + d^2/16)}{g\,a/2}}$$

If d is small compared to a,

$$T = 2\,\pi\sqrt{2a/3g}$$

(If d is less than 0.07 a, the error in g through neglecting d will be less than 0.1 per cent. You want greater precision than this.) Measure a, d, and T; and calculate g.

2. The form of pendulum illustrated in Figure 8-4 **B** is a simplified variation of what is called the Kater pendulum. It consists of a rectangular metal bar oscillating about a support at a special distance from the center, this distance being known as the radius of gyration. If the bar is supported near the center, the period will be very long, and as the support is moved out the period decreases. At a distance equal to the radius of gyration the period is a minimum, and beyond it the period increases again. The advantage in using the radius of gyration as a point of support is that the period changes only very slightly if the support is not exactly at that point. Any error in the period resulting from the support's not being at exactly that position is very small, and that is what we are after, small errors.

The moment of inertia of a rectangular bar about the center of mass is

$$I_{cm} = m(a^2 + b^2)/12$$

If the mass m were concentrated at a radius r, the moment of inertia would be

$$I = mr^2$$

If r is chosen to be $(a^2 + b^2)/12$, this would be equal to I_{cm}. The quantity r is an "effective" radius for the rotation of the bar and is called the radius of gyration. The length, a, and the width, b, of the bar can be measured precisely. For a, use two measuring microscopes and a standard meter bar as described. Then r can be calculated and the support placed as close as possible to this distance from the center of the bar (the balance point).

Following the same line of development of the expression for the period as in the previous section,

$$I = 2mr^2 \quad \text{(this may require some thought)}$$

$$L = mgr \sin \theta$$

$$= mgr, \text{ if } \theta \text{ is small}$$

so

$$K = mgr$$

then

$$T = 2\pi\sqrt{2mr^2/mgr} = 2\pi\sqrt{2r/g}$$

r has been calculated from measured values of a and b. Determine T to a sufficient accuracy. Then calculate g and the error in g based on the estimated uncertainties in the measurements.

3. A hoop with an inside radius of r, an outside radius of R, and a mass of m. (See Figure 8–4C.) This type of pendulum is very good for obtaining the precision required in this project. The hoop can be machined from a steel pipe, and although the size is not important, a 10-inch outside diameter pipe with a quarter-inch wall will yield a pendulum that has a period very close to one second. The timing can then be done using the one-second beats such as are put out by WWV or a similar station.

$$I_{cm} = m (R^2 + r^2)/2$$

$$I = m (R2 + 3r^2)/2$$

$$L = mgr \sin \theta$$

$$= mgr \, \theta \text{ for small values of } \theta$$

$$k = mgr$$

$$T = 2\pi\sqrt{m(R^2 + 3r^2)/2mgr}$$

$$= 2\pi\sqrt{(R^2 + 3r^2)/2gr}$$

Measure r, R, and T, and solve for g. Calculate the error in g.
If the hoop is thin so that we can let $R = r + \Delta r$, you can show that the expression for the period reduces to

$$T = 2\pi\sqrt{(2r + \Delta r)/g}$$

The inside radius is r and the thickness of the hoop in the radial direction is Δr. Is this approximation sufficiently precise?

Remember that this is a project in precision measurement. The error calculation is important, and the error in each quantity must be kept to a minimum. Your final result, with its error, should be compared with the theoretical value for your altitude and latitude as given by the formula on page 146.

Apparatus

One form of physical pendulum with a solid knife edge support

Calibrated stop watch (or short wave receiver capable of receiving a standard time broadcast)

Standard meter scale with lenses or measuring microscopes

Thermometer

8-4

Torsion Pendulum

A torsion pendulum is one which has a twisting motion, rather than the back and forth motion of the simple pendulum. Torsion-type devices, like the torsion pendulum, are used in some very sensitive instruments to measure such things as magnetism, electric current, and electric charge. The use of one of these devices to measure the gravitational force between two masses is described in a later experiment. This project, in fact, may be regarded as preparation for the measurement of the gravitation constant.

Such sensitive measuring instruments measure small forces by using them to twist a fiber or fine wire. The torque, L, to twist a wire through an angle θ is described by $L = k\theta$. k is called the torsion constant and is what must be determined in a measuring device. The torsion constant may be found by using the dimensions and elasticity of the suspension, by the direct application of a known torque, or by using the wire in a torsion pendulum. The purpose of this project is to find the torsion constant of a wire by all three methods.

Reference to a physics text will show that theoretically the torque required to twist a solid rod (a wire or fiber is a very thin rod) through an angle θ in radians is described by

$$L = \frac{E_s \pi r^4}{2 l} \theta$$

where r is the radius, l is the length of the rod, and E_s is the shear modulus of the material. For steel the shear modulus is about 8×10^{10} newtons per square meter. This value will vary somewhat from one sample of steel to another, but you can use it to calculate an approximate value of the torsion constant. You could allow a 20 per cent uncertainty.

With the disc provided fastened to the wire, use the arrangement shown in Figure 8-7 to apply a torque to twist the wire through one revolution. Then find the torsion constant.

For the third method to determine the constant, oscillate the disc as a torsion pendulum and measure the period. In the case of a force described by $F = -kx$, the period of linear vibration is given by $T = 2\pi\sqrt{m/k}$. Similarly, if the restoring torque is described by $L = -k\theta$, the period for angular oscillation will be described by $T = 2\pi\sqrt{I/k}$. I is the moment of inertia of the oscillating object. If you use a solid disc of mass M and radius R, then $I = MR^2/2$. Make the appropriate measurements to determine k by this method. Is the agreement of the three values of k satisfactory?

You will note that the torsion pendulum could be used to measure the shear modulus or to find the moment of inertia of an object fastened to it.

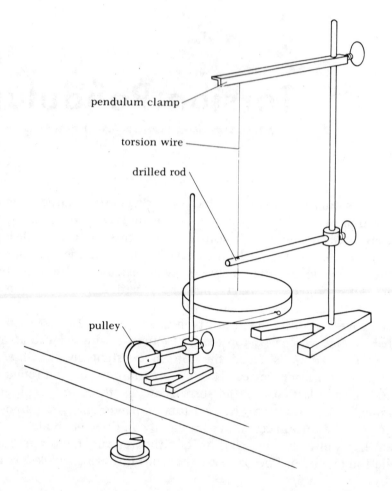

Figure 8-7 Obtaining the torsion constant of a wire by the direct application of torque.

Apparatus

Meter scale
Micrometer caliper
Torsion wire
Lab stand and pendulum clamp
Metal disc to fit on wire
Pulley and holder
Weights and weight holder
Stop watch
Balance

8-5

Young's Modulus

You are provided with a sturdy stand from which hangs a wire. On the end of the wire is a weight holder as illustrated in Figure 8-8. As you put on and take off weights nothing extraordinary seems to happen. However, watch the bottom of the wire through a microscope as a weight is put on and then taken off! Put on two or three weights while you watch through the microscope and then take them off, again while looking through the microscope. You actually see the stretching of the wire, and the wire goes back to the original length when the weight is taken off. Not only can you see the elongation, but with the microscope provided you can measure it.

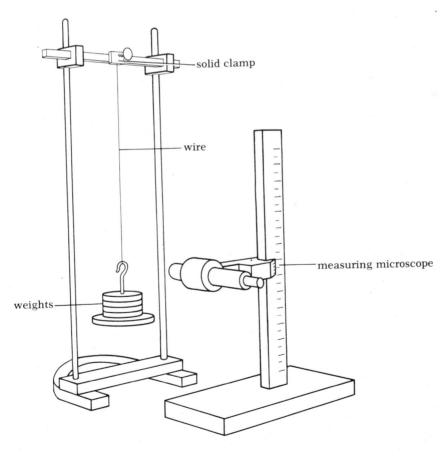

Figure 8-8 The apparatus used to see and to measure the elongation of a wire when a force is applied to it.

In case the knot or twist on the bottom of the wire slips, it is advisable to put a small mark on the wire near the bottom and to use it for measurement. Also, the elongation may be only partly due to stretching and partly due to bending of the support. The microscope can be used to measure this, and take it into account if necessary.

The wire is actually elastic and you are to determine how elastic the wire is. The measure of this is the quantity called *the modulus of elasticity*, which, for linear deformation such as this, is known as Young's modulus.

The modulus of elasticity (look it up in a text) is defined as stress over strain, where stress is force per unit cross-sectional area (F/A) and strain is elongation per unit length (e/l) caused by stress.

$$Y = \frac{F/A}{e/l} = \frac{Fl}{Ae}$$

Make the necessary measurements to get as accurate a value of Y as you can. Compare your value with a value listed in tables. Such a comparison serves mainly to show that your value is a reasonable one, but *the actual modulus for that wire is the one that you measured.*

The error in your final value should be included, and the simplest way to find this is to express Y in terms of the measured quantities. For example, the diameter is measured in order to determine A because ($A = \pi d^2/4$), and the force is mg. Substitute these expressions in the expression for Y and then use this single expression in finding the error in Y. (See Chapter Three for the calculation of error.)

Apparatus

Strong support for wire
Wire
Measuring microscope
Meter scale
Set of weights and weight holder
Micrometer caliper

Prism Spectrometer

The object of this experiment is to determine the index of the glass used in a prism. A spectrometer will be used to make an accurate measurement of a prism angle, A, and then to measure the angle by which light is deviated by the prism. For the angle of minimum deviation, D, the index of refraction is given by

$$\mu = \frac{\sin \dfrac{A + D}{2}}{\sin \dfrac{A}{2}}$$

Because the index varies with color, white light is dispersed by the prism. The apparatus for this experiment is sufficiently accurate that you can measure the index of refraction for whatever colors you wish, and you should do it for at least two colors.

For the operation and focusing of a spectroscope refer to Chapter Ten.

The prism angle, A, is nominally 60 degrees but may differ from this by a degree, so it must be measured. To measure this angle, turn the apex of the prism toward the collimator and set the telescope to read the angle at which light is reflected off each face as illustrated by two of the telescope positions of Figure 8-9. You may show that the angle between these rays as shown in the diagram is

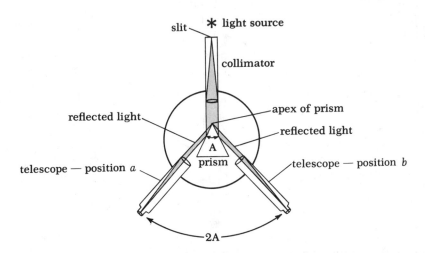

Figure 8-9 The use of a spectrometer to measure the angle of a prism. The apex is turned toward the collimator and the direction of the light reflected from the two faces is measured by setting the telescope in each of the two positions shown.

equal to 2*A*. In some way mark the angle which was measured. It is easy to measure one angle and then use another unless the prism has a frosted base or unless the angle was marked by a small piece of tape or ink mark on the top of the prism.

Measure the angle of minimum deviation as follows: When a light ray goes through a prism, its direction is changed by an amount depending upon the angle at which it hits the prism as well as upon the prism angle and the index of refraction of the prism. With the prism on the prism table, oriented as shown in Figure 8–10A, and with the light in front of the slit, set the telescope to see the light coming through the prism. A spectrum ranging from red to violet should be seen. Then turn the prism with your fingers until the angle through which the light is deviated is as small as you can make it. Set the cross hairs on the color for which you wish to measure the index μ and read the angle (using the vernier scale to measure to 0.1 degree). To find the angle of deviation, remove the prism and read the angle when the telescope is set on the light coming straight through as shown in Figure 8–10B. Calculate the angle of deviation. Alternatively, reverse the prism so that the deviation is on the opposite side and the total angle will be 2*D*. This is illustrated in Figure 8–10C.

The result of this experiment can be presented in a table showing the index of refraction for light of various colors. From reference to texts or tables giving representative values for different types of glass, the kind of glass from which the prism is made could be deduced.

An estimate should be made of the precision of the calculated values of the index. The actual calculation of the error from your estimates of how precise your measurements of *A* and *D* were requires that you use the method of partial

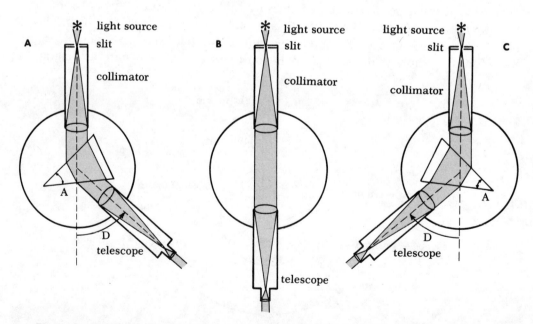

Figure 8–10 The spectrometer as used to measure the angle of minimum deviation of light by a prism. The prism is reversed in **C** from what it is in **A** and the angle of deviation, *D*, is half the angle between the two telescope positions shown. In **B** the telescope is shown set in the position to measure the direction of the undeviated light.

differentiation as described in Chapter 3. This is necessary not only because trig functions are involved, but also because the angle A occurs in both the numerator and the denominator, so the effect of an error in A is reduced. You should carry this through to see if the following expression for the relative error is correct. It has been adjusted for errors in A and D to be expressed in degrees.

$$\frac{d\mu}{\mu} = \frac{1}{114.6} \left\{ \left(\cotan\left(\frac{A}{2}\right) - \cotan\left(\frac{A+D}{2}\right) \right) dA + \cotan\left(\frac{A+D}{2}\right) dD \right\}$$

You may be impressed by the precision of your result.

Apparatus

Spectroscope
Incandescent lamp
Prism

8-7

The Universal Gravitation Constant, G

One of the important fundamental constants in physics is the gravitation constant G, which occurs in the expression of the law of gravity in the form $F = G\, m_1 m_2 / r^2$. F is the force between particles of mass m_1 and m_2, separated a distance, r. The masses involved are point masses, but the formula also holds for spherical masses. One of the first applications of calculus, in fact probably a motivation for its development, was in showing that in the case of spherical objects the distance r could be measured between the centers of the spheres.

The gravitational force between objects in the laboratory is very small and the apparatus used to measure it, therefore, is very sensitive and delicate. The measurement was first made by Henry Cavendish in 1797–1798, and there has been little improvement in his basic method since. Cavendish's report is readily available in *Great Experiments in Physics,* a series of original works collected by M. H. Shamos.[2] You will use a torsion balance similar to that used by Cavendish and many others, even by those in today's modern research laboratories. The apparatus is illustrated in Figure 8-11.

If the torsion constant of the fiber and the displacement of the small balls when they are attracted by the large ones are measured, then the force can be calculated. The masses of the balls and various dimensions must be known in order to calculate G.

The large balls can be weighed on a balance, but it has been mentioned that the apparatus is very delicate, so you must not open the case to weigh or to measure the small balls. The fiber is very fine, easily broken, but not easily repaired. The masses of the small balls and the internal dimensions of the apparatus used may be supplied by the instructor. Or the measurements of the apparatus can be made from the outside using a measuring microscope and the masses of the small balls can be calculated from their size, using the density of lead as found in tables. The other quantities to be found are the torsion constant (deduced from the period of oscillation and the calculated moment of inertia) and the change in equilibrium position of the small balls when the large ones are in the positions shown in **A** and in **B** of Figure 8-12. The angles are measured by reflecting light onto a scale from the curved mirror which is attached to the suspension. When the mirror turns through an angle θ, the reflected light changes direction by an angle 2θ. You should verify this to yourself.

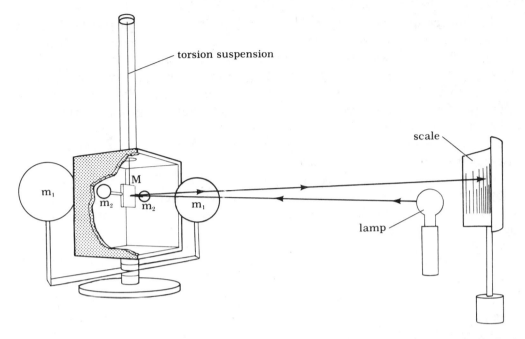

Figure 8-11 The Cavendish apparatus to measure the force of attraction between two lead balls.

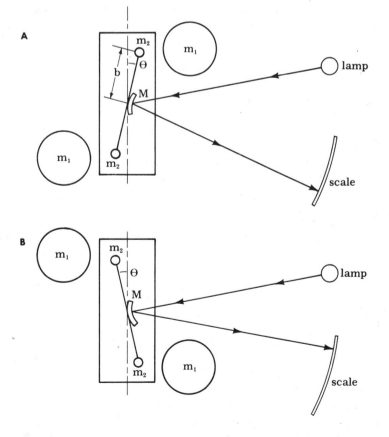

Figure 8-12 A diagram to show the use and analysis of the Cavendish apparatus.

The procedure for the experiment is as follows: First let the small balls come to rest with the large balls in the position shown in **A** of Figure 8-12. Note the position of the light on the scale. Carefully move the balls to the position shown in **B** and start a stop watch. Record the position of the light spot on the scale every minute for 30 minutes then every two minutes for another 30 minutes. After the first ten minutes start plotting a graph of the position of the spot against time. From the graph determine the period of oscillation and the second equilibrium position. To find the angular displacement, you must also know the distance from the mirror to the scale.

The torsion constant, k, is related to the period, T, and the moment of inertia, I, by $T = 2\pi\sqrt{I/k}$. The moment of inertia of two spheres, each of mass m and radius a and with their centers at a distance b from the axis of rotation, is given by

$$I = 2m(b^2 + 2a^2/5)$$

The period is obtained from the graph of the position of the light spot against time.

The angular displacement between the positions of the light on the scale with the large balls in the positions shown in Figure 8-12A and **B** is four times the angular displacement that would be obtained for attraction between only one large and one small ball. The reason for this is that the attraction is first in one direction and then in the other and also that turning a light beam moving through an angle θ results in the direction of the reflected light beam moving through twice that angle. The angle is calculated from the distance the light spot moves and the distance from the scale to the mirror.

Apparatus

Cavendish apparatus (with a suitable scale and light source) for the measurement of G.
Clock with sweep second hand (or stop watch)
Balance suitable for weighing the large lead balls
Meter scale
Measuring microscope

Planck's Constant

The modern particle theory of light started with Max Planck in 1900 when he found that in the theoretical analysis of radiation from a so-called black body he had to assume that light existed in "quanta," each having an energy given by $E = hc/\lambda$, where c is the speed of light, λ is what we measure as the wavelength, and h is a proportionality constant which we now call Planck's constant. Since that time Planck's constant has been found to appear in many situations in modern physics. It is one of the fundamental constants.

In 1905 Einstein showed how to measure Planck's constant experimentally and it is his method, using the photoelectric effect, that you will follow. The photoelectric effect is dealt with in Chapter Ten; you should read the appropriate part.

When a "particle of light" or photon, hits an electron, the energy may be transferred to the electron, so the idea is to measure the energy of electrons which are knocked out of a material by light, as occurs in a photocell. Since, as described in Chapter Ten, the photoelectric effect occurs only with bound electrons, some of their energy is lost before they come out of the material. The electrons are emitted from a substance with a range of energies up to a maximum which is less than the energy of the photon by an amount W, called the work function of the material. W is different for different materials. So the maximum energy of electrons ejected by light of wavelength λ is given by

$$E_{(electron)} = E_{(photon)} - W$$
or
$$E_e = hc/\lambda - W$$

W can be eliminated by measuring the maximum electron energy at each of two different wavelengths, λ_1 and λ_2, and then subtracting. You can carry out the analysis to solve for h if the wavelengths and the velocity of light are known and if the two electron energies are measured.

You will use a vacuum type photocell, which is also described in Chapter Ten.

To measure the electron energy, make use of a fundamental definition in electrical units; that is, when 1 coulomb of charge moves across 1 volt, it acquires an energy of 1 joule. If a charge e is attracted from one plate to another by a voltage V, it acquires an energy eV joules. One electronic charge (1.602×10^{-19} coulombs) moving across 1 volt gains 1.602×10^{-19} joules which is used as a unit of energy called 1 electron volt eV.

In this experiment the electrons will leave the photosensitive surface with an energy E and the collecting rod, which would ordinarily be charged positively to collect them, will be made negative to stop them. When the negative voltage V is just sufficient to stop them, $E = eV$, so you have a method to measure the energy E. As we pointed out, the electrons have energies from zero up to a certain maximum. To find the maximum, measure the photocell current at different increasingly negative voltages on the anode and plot a graph of current against stopping voltage. Then extrapolate it to find what voltage would stop them all. If an increase in voltage increases the current rather than decreases it, it means that you have made the anode positive rather than negative as you must do in this experiment.

To get two wavelengths use two filters. The wavelength passed by the filter will be given to you by the instructor. Find the maximum energy of the electrons ejected by the light from each filter and then find Planck's constant.

The wiring diagram is shown in Figure 8-13. The voltage source V, need be only a dry cell and the rheostat should have about 1000 ohms. The meters used are more sensitive than usual.

To perform the experiment, mount the photocell and the lamp provided on a stand with a holder between for the filter. Adjust the light position to give a current reading near full scale with a filter in position and with no voltage on the cell. It will be necessary to use a very bright lamp and perhaps a lens to focus it onto the photocell in order to obtain sufficient current with the filters in place. It may then be convenient to switch it on only briefly while each reading is obtained.

Once you have found Planck's constant you can find the energy in 1 quantum of the light of each of the colors you used.

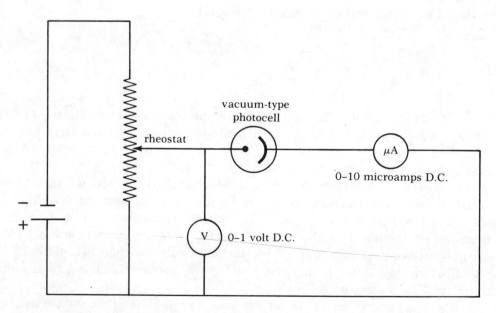

Figure 8-13 The wiring diagram for the experiment to determine Planck's constant.

Apparatus

Vacuum type photocell
1–1.5 volt battery
Rheostat (about 1000 ohms)
D.C. voltmeter–0-1 or 0-1.5 volts
D.C. microammeter–0-10 μA
Filters of known wavelength
200 or 500 watt lamp

REFERENCES

1. Childs, W. H. J.: Physical Constants Selected for Students, Ed. 5. London, Methuen and Co. Ltd., 1949, p. 7.
2. Shamos, M. H.: Great Experiments in Physics. New York, Henry Holt and Co., 1959, p. 75.

9

PHYSICAL ANALYSIS

One of the applications of physics in this modern day is the analysis of materials by what could be called physical methods, as contrasted to chemical methods. Some physical methods require destruction of a portion of a sample, but others do not; and testing by means of the latter methods is referred to as *nondestructive testing*. Physical analysis methods may draw from almost any branch of physics.

Of the historic examples of nondestructive testing, perhaps the best known is Archimedes' analysis of the composition of a crown. The story goes that the king had reason to suspect that the smith who made the crown had kept some of the gold provided and had made up the amount by alloying the remaining gold with some other metal. The finished crown was the same weight as the gold provided for its construction, but how could one tell if the crown were pure gold? Archimedes' solution was to find the density by the principle we now attribute to him. The method is outlined in Experiment 9-1. The outcome in this case was that the density was less than for pure gold and the goldsmith was found guilty of stealing some of the precious metal provided to him.

A more recent example is in connection with investigation about the circumstances of the death of Napoleon. One theory was that he had been slowly poisoned by the regular addition of small amounts of arsenic to his food. Some hairs from his body were irradiated in a nuclear reactor and the characteristics of the radioactivity induced in the hair were analyzed. The radiation was that which could have been produced only by the presence of arsenic in the hair. The amount of arsenic had been demonstrated; the rest of the story is to be found by the historical investigators.[1]

For those using irradiation facilities (nuclear reactors or accelerators) to study nuclear reactions, the fact that exceedingly small amounts of some materials can be detected is very troublesome, for the purity of materials required in investigations is very high. Exceedingly small amounts of impurities may mask an effect being studied. It is odd (or is it?) that what is troublesome in one area provides a valuable tool to do something in another area. There seems to be "good" or "bad" in everything. It depends on the situation at hand.

X-rays and gamma rays provide another tool for nondestructive testing. Metal castings are often radiographed, using X-rays or gamma rays, to detect flaws in the form of cracks or bubbles. Welds are often checked in the same way. Every bit of welding on such important pieces of equipment as the containment vessel for a nuclear reactor is radiographed because the finished product must be flawless.

The spectroscope and variations of it are also used in analysis. The best known method is the analysis for elements by inserting a sample in an arc and then observing the spectral lines. Each element has its own pattern of lines which allows its presence to be identified. Such analysis is routine in many industrial, crime, or research laboratories. Similarly, the presence of materials, even complex molecules, in solutions can be identified by the color of the light absorbed by that solution. Instruments designed for this purpose usually measure also the amount of absorption at different colors, indicating also the amount of material present. Such instruments usually use photocells of one type or another to measure the light, and they are called *spectrophotometers*.

Wavelengths of spectral lines associated with various elements are listed in Table 9–1. These are arranged according to approximate color, and some of the ultraviolet lines are included. Also, only the brightest are shown. Those enclosed in parentheses are so close together that they may appear as one, and those marked *B* refer to the edge of a band rather than a sharp line. The wavelengths are in nanometers (10^{-9} meters). Table 9-1 can be used in conjunction with Table 9–4 for the experiments on spectral analysis. The column marked Hydrogen Balmer indicates a set of lines that are emitted by hydrogen under special conditions. They are included partly because of their occurrence in stars, and one of your optional projects may involve the analysis of the spectrum of a star.

Some materials, when dissolved in water, change the index of refraction—the index depending on the concentration of the dissolved substance. Instruments designed to analyze solutions by the measurement of the index of refraction are called *refractometers*. Sugar solutions are often analyzed in this way.

Another property used in analysis is the rotation of polarized light. When polarized light passes through some solutions, the direction of polarization gradually rotates. The amount of rotation in a given path depends on the amount of material. Again sugar is an example. Sugar solutions rotate the plane of polarization of light, and *polarimeters* are commonly used to measure sugar concentrations. If we looked toward the light beam, the rotation could be counterclockwise (left) or clockwise (right). Some sugars can be referred to as "left-handed" (levulose) and others as "right-handed" (dextrose), although these descriptions are not in general use. Most people would question your sanity if you insisted that sugar may be right- or left-handed. These properties, the rotation of the plane of polarization and the direction of such rotation, are rather interesting to investigate theoretically. Although such investigations are left for more advanced courses, the use of these phenomena for analysis is possible knowing just that they exist.

TABLE 9-1 The Wavelengths of Spectral Lines from Some of the Elements. The units are nanometers. Only the brightest are shown, and, therefore, the table could not be called complete.

Approximate Color	Hydrogen H	Hydrogen Balmer	Helium He	Neon Ne	Bromine Br	Krypton Kr	Mercury Hg
			668	644			691
Red		651		641			
				634			
				626			629
	612			615	612B		
Orange				607			
	603			603			
Yellow	598		588	598	594B		
	581			588		587	579
				585			577
Green				540	559	558	546
			502		518		
Blue	493	486	492		482		492
	463		471		453B	446	
						445	
		434	439		437	436	436
						427	
Violet		410	412			414	
			403				405
Ultraviolet		397					
		389					
		384					366
		380					

9-1

Analysis by Density or Specific Gravity

A simple but careful density or specific gravity measurement can provide considerable information about the composition of an object. If a coin is supposedly made of silver, and even has the look of silver, the silver may, in fact, be alloyed with another metal, and in most cases will be alloyed with copper. Pure silver has a density of 10.50 gm/cm^3, and copper only 8.94 gm/cm^3. An alloy would have a density between these, the actual value depending on the fraction of copper present. It is apparent also that to conduct even an approximate investigation of the composition, careful and precise work would have to be done. Density measurements do not provide the same kind of information as a chemical analysis. For example, density measurements do not reveal that the alloying metal is in fact copper, but they do give some definite information and, further, are in the category of *nondestructive testing.*

One can refer to the density and specific gravity of a *material*, and these are quite definite values that are tabulated in various handbooks. Also the density and specific gravity of an *object* can be measured. These will be average values, and they depend on the composition and structure of the object.

The terms *density* and *specific gravity* should be clarified. Density refers to mass per unit volume. The units may be of a great variety: gm/cm^3, lb/ft^3, kg/liter, kg/m^3, etc. Specific gravity, s.g., is the weight of an object (pure substance or not) compared with the weight of an equal volume of water. You could show that this is equivalent to the ratio of the density of the object (or material) to the density of water. Because the s.g. is a ratio, it has no units and the value does not depend on whether the measurements were done in the English or in the metric system. In the metric system, however, the density of water is 1 gm/cm^3 or 1 kg/liter, and the density and s.g. are numerically the same.

For irregular objects the s.g. is found most readily by using Archimedes' principle. According to this principle, the buoyant force on an object immersed in water is equal to the weight of the water displaced. The buoyant force shows as a loss in weight when the object is immersed in a fluid. So to find the s.g., weigh the object in air and then suspend it by means of a thread from the hook on the balance as shown in Figure 9-1 to weigh it when it is immersed in water. The s.g. is the object's weight in air compared with the loss of weight when it is immersed in water. You will use a balance weighing to a hundredth of a gram, and don't forget to consider the weight of the thread.

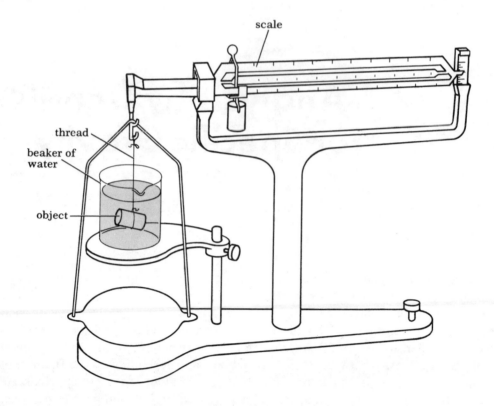

Figure 9-1 The use of a balance to weigh an object in water.

The determination of s.g. in this way requires that you work extremely carefully. It is an exercise in care and precision. Your results will be a measure of your effort, and if they have achieved the goal, you may be justly proud.

Some suggested analyses follow:

A.   Analysis of the metal of a coin

The analysis of a coin has been suggested. If you have a supposedly silver coin, find the s.g. and make an estimate of the amount of copper in it. Estimate the error in each measurement, in the s.g. and in the estimated composition.

B.   Identification of metals or other materials

You will be provided with some metal samples, wires, blocks or rods, and you are to make a probable identification by the measurement of the s.g. For example, one of them may be very light, and you suspect that it is aluminum. However, for aircraft use aluminum is often alloyed with the even lighter magnesium. Is your sample pure aluminum or possibly an alloy with magnesium, and if so, what fraction is magnesium?

Lead of s.g. 11.34 is often used for shielding against gamma rays. Lead, however, is very soft, and often it is alloyed with antimony (s.g. 6.62) to harden it, to make it easier to work with and to make it less liable to become distorted. Is your sample heavy? Is it lead or an alloy?

In Table 9-2 is a list of various materials with their specific gravities. Your project may involve one or more of these, but the basic ideas are illustrated above.

C.   Analysis of earth, planets, etc.

TABLE 9-2   The s.g. of Various Materials.

Material	s.g.
Aluminum	2.71
Magnesium	1.74
Lead	11.34
Antimony	6.62
Silver	10.50
Copper	8.94

Astronomers (with the help of physicists) have deduced that the average density of the earth is 5.52 times that of water; that is, the average s.g. is 5.52. Some other calculated average specific gravities are shown in Table 9-3.

The object of this project is to make some measurements of the s.g. of a variety of samples of the earth's crust (rocks), and then make some speculations about the composition of each of the objects in Table 9-3. Find also the s.g. of cheese (green, preferably) to show a necessary change in an ancient myth about the moon.

TABLE 9-3   The Mean s.g. of Some of the
Objects in the Solar System.

Object	s.g. or Density in gm./cm.^{3}
Earth	5.52
Moon	3.4
Mercury	5.1
Venus	5.3
Mars	3.94
Jupiter	1.33
Saturn	0.69
Sun	1.39

Apparatus

Balance weighing to 0.01 gm., with a hook and stand for use with Archimedes' principle

Thread

Objects of which the s.g. is to be found

silver coins

rocks (and cheese)

aluminum or aluminum alloy

lead or lead alloy

9-2

Spectroscopic Analysis

The objects of this project are to identify the element in a given light source by measuring the wavelengths of the lines in the spectrum of the light emitted and also to find the wavelength range of light that you can see.

The wavelengths are to be measured with a diffraction grating spectrometer, as shown in Chapter Ten. The instrument must be focused as described in that chapter and a diffraction grating of about 15,000 lines per inch is put on the prism table. The principle of the diffraction grating is also described in Chapter Ten. The grating must be placed as nearly perpendicular as possible to the light beam from the collimator. When the adjustments are properly made, the room lights dimmed, and the spectrometer pointed toward an incandescent lamp, the spectra will be seen by your looking into the telescope and moving it on either side of center. At least two complete spectra will be seen on each side of the bright white line in the center. The focusing should be such that the center line is very sharp and the slit of the collimator narrowed, so that the white center line appears as only a thin line.

While the continuous spectrum is being observed, the wavelengths at the red and at the violet end could be measured. You can then see to what range of wavelengths your eye is sensitive. This range is often said to be from 400 nm to 700 nm, but this is a crude approximation. You have the opportunity here to measure it for your own eye, and even more precisely than that; you could measure the wavelengths associated with the various colors that you see. The procedure for finding the wavelengths is outlined below.

For element identification, the incandescent light is replaced with a gaseous discharge tube. Instead of continuous spectra, bright spectral lines will be seen. You identify the element in the tube by measuring the wavelengths of those lines.

The first spectrum on each side of center is called *first order* and has $n = 1$ in the relation $n\lambda = s \sin \theta$. (See Chapter Ten.) For the second spectrum $n = 2$, etc. To calculate the wavelengths, the relation $\lambda = (s \sin \theta)/n$ is to be used. The line spacing, s, is calculated from the given number of lines per unit length across the grating. This is ordinarily marked on the grating. This spacing should be calculated and expressed in the unit desired for the wavelength, which is millimicrons (mμ) or nanometers (nm.), each of which are expressions for 10^{-9} meters.

If the grating is not exactly perpendicular to the light beam, the angles to the same line on the right and left will be slightly different. The average will be better than either. A quick way to find this is to measure the total angle between the lines to the left and right and divide by 2. A systematic way to acquire the data is to start at 90 degrees from the central beam and move the telescope in just one direction, stopping to set it on each line encountered and recording the color, brightness (descriptive), and protractor setting for each. All observations should be recorded in a table.

In the table on the report sheet, 2θ is the angle between the protractor readings on the same line on the right and left. The columns 2θ, θ, $s\sin\theta$, and λ are all calculated. Possibly only the brightest lines will show in the second order, and some may even show in the third.

Lists of some of the wavelengths associated with a few of the elements are given in Table 9-1. The wavelengths are in millimicrons or nanometers.

If your pattern of wavelengths does not correspond with one of those in Table 9-1 or 9-4, check your measurements, and if you still disagree, you may have to look up other elements in the library. However, consult with the instructor first, for air may have leaked into the tube.

Apparatus

Spectrometer
Grating–15,000 lines per inch
Light bulb
Discharge lamp –Hg, Ne, He, H, etc. (but one which is included in Table 9-1)

9-3

Spectral Analysis with a Camera

In this project a spectrograph consisting of a camera and diffraction grating is used to identify elements in glowing light sources. The sources will be discharge tubes in the laboratory, but the equipment can also be used to identify elements in distant street lights and even in some of the brighter stars. These latter are optional extensions of the project.

To transform the camera to a spectrograph, a grating of about 15,000 lines per inch is placed in front of the lens. The action of the grating is described in Chapter Ten and you may also find it in another text. An example of a photograph of four discharge tubes, one above the other, is shown in Figure 9-2. The discharge tubes are imaged in the center, and on either side of each is a spectrum.

The distance on the film from the central image to each line depends on the wavelength. If the wavelength of one of the elements in the photograph is known, a graph of wavelength against the distance from the central image of the source may be constructed; this will be the calibration curve for that photograph. The wavelengths of the unknown spectra are found by measuring the distance from the center to each spectral line and then using the calibration curve. Careful measurement on the photograph using a traveling microscope accurate to 0.01 mm will give wavelength determinations to the nearest nanometer or so. This precision exceeds that obtainable with a spectrometer, which reads only to 0.1 degree. A further advantage of the photographic method is that some of the

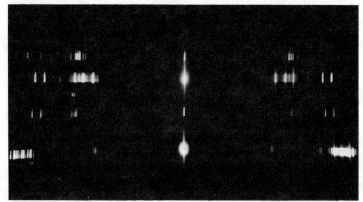

Figure 9-2 Four discharge tubes in a darkened room photographed through a diffraction grating. In the center are the images of the tubes and on either side are the first order spectra. Mercury is in the tube second from the bottom.

ultraviolet rays will pass through the lens and affect the film. Spectral lines in the near U.V. are therefore included in this experiment.

The light sources are placed on shelves, one above the other, with a plywood light shield in front of them and slits to allow viewing of the sources, one of which must be the known. If the room lights are turned off and the array viewed through a grating held close to the eye, spectra of each will be seen on either side. This is what will appear on the photograph. With black and white Polaroid film of speed 3000, an exposure equivalent to 1 second at f/5.6 will probably give satisfactory results. A distance of about 4 meters was used for Figure 9-2.

Mercury may be used as a standard, and in Figure 9–2 the second spectrum from the bottom was mercury. The principal wavelengths in the mercury spectrum are shown in Tables 9–1 and 9–4. An approximate calculation will allow them to be identified, and then the accepted wavelengths from Tables 9-1 and 9-4 can be used in the making of the calibration curve. To make the approximate wavelength calculation, the angle between the central maximum and each line must be found. This is the angle shown as θ in Figure 9–3. The distance x is measured, and from the known lens-to-film distance (use the focal length marked on the camera and the measured lens-to-object distance to calculate this) the angle θ is found. The grating equation $\lambda = s \sin \theta$ is used to calculate λ. The distance between adjacent grating lines is s, and it is found from the number of lines per unit distance, which is marked on the grating. Once the standard lines are

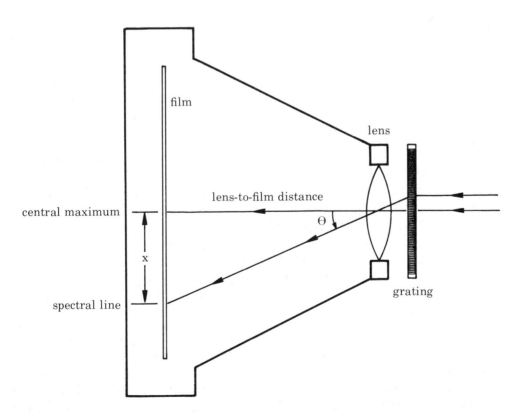

Figure 9–3 A diagram to assist in the calculation of the wavelengths of the spectral lines on a film. The angle θ is the angle occurring in the grating equation, $\lambda = s \sin \theta$.

TABLE 9-4   The Wavelengths of the Spectral Lines of Some Elements.
This table is supplementary to Table 9-1. The wavelengths are in nanometers,
and only the brightest are shown.

Mercury	Hydrogen	Helium
366.3	379.8	388.9
404.7	383.5	396.5
435.8	388.9	402.6
546.1	397.0	412.1
577.0	410.2	438.8
579.1	434.0	447.1
	486.1	471.3
	656.8	492.2
		501.6
		504.8
		587.6

identified, the calibration curve can be made. The distance x for each line in the unknown spectra can be measured, the wavelength found, and the sources identified. This is the main purpose of the project.

Further work:
The camera and grating as described can be used to obtain other spectra as follows:

1. Either outdoors at night or through a window, the spectrum of a vapor-type street light may be obtained for identification of the element producing the light. A discharge tube from the lab should be included in the photograph to make a calibration curve.

2. The camera and grating may be mounted on an equatorial telescope mount and pointed toward a bright star. A tube or long open box in front of the camera may be used to eliminate light from other objects. An exposure of 3 or 4 minutes on film of speed 3000 will give satisfactory spectra of bright stars such as Sirius, Rigel and Vega. The element responsible for the lines in those spectra can then be identified.

To obtain a spectrum of a star, the camera is kept stationary, and during the exposure the star moves, leaving a track on the film and spreading the spectrum into a wide band. A mercury discharge tube placed at a distance and photographed, on either the same film or another, will be a standard.

3. A solar spectrum can be obtained by putting an opaque board with a narrow slit in it over a window that looks toward the sky. This slit can be photographed from about 4 meters, with the grating in front of the lens. An exposure of a second or so will yield a spectrum crossed by the dark lines, which can then be identified. These dark lines in the solar spectrum are called *Fraunhofer lines*. A mercury discharge tube can be put above or below the slit for a standard.

4. The spectra of planets such as Jupiter, Saturn and Venus can be obtained in the same way as the stellar spectra outlined above. The spectra of the light from the planets can then be compared with the solar spectrum obtained in 3.

Apparatus

Polaroid camera and film of Speed 3000
Four discharge tubes with power supplies
Box with shelves and slits to hold tubes and power supplies
Diffraction grating, 15,000 lines/inch, good quality student-type
Tape to fasten grating to camera
Black cloth to eliminate stray light from the box
Measuring microscope
For optional parts
    board with slit to cover window
    equatorial-type telescope mount (not motor driven)
    mount with tube or open box to hold camera on telescope mount
    mercury discharge tube

REFERENCES

1. Forshufvud, S., Smith, H., Wassen, A.: Arsenic content of Napoleon I's hair probably taken immediately after his death. Nature. *192*:103 (1961).
2. Greenberg, L., and Balez, T.: Spectral analysis using a camera. American Journal of Physics. *40*:319 (1972)

10

MEASURING INSTRUMENTS

In physics laboratories a great variety of equipment is used — in comparison to, say, a biology lab where most of the work can be performed with a microscope. When the student masters this item, his effort can then be directed toward the work being done with it. He need not worry about the manipulation of the instrument.

Instruments are partly a means to extend our senses beyond the ordinary range or to an area in which our senses do not operate: Microscopes are used to extend our vision; electric meters measure something which does not affect our sensing organs. In physics the instruments also have the important purpose of making measurements.

Some of the instruments you will be using in the laboratory are described in this chapter. The principles of the instruments and the theory behind their operation are dealt with. In some cases details of actual instruments are given, but you may encounter some instruments which are not dealt with here. If instruction sheets are supplied with a new instrument, *be sure to read them* so that you will understand the equipment and will then be able to use it to obtain correct measurements, without damaging the mechansim.

MEASUREMENT OF LENGTH

Three instruments used for the measurement of length will be considered. These are the vernier caliper, the micrometer or screw caliper, and the measuring microscope. The use of light waves for measurement and the determination of wavelengths will also be discussed.

THE VERNIER CALIPER

The vernier principle which allows accurate determination to fractions of the divisions on the main scale is used in a multitude of instruments. The vernier caliper is illustrated in Figure 10–1. The diagram labeling is self-explanatory, but the way in which readings are made is as follows: (Only the metric scale will be explained.) The vernier scale is the small one that moves along the main scale. With the caliper closed, the zero end of the vernier coincides with the zero mark on the main scale. With the caliper open by 1 mm. the zero on the vernier lines up with the 1 mm. mark on the main scale, and so on. The millimeter reading is obtained from the zero end of the vernier scale. To see the operation of the vernier, close the calipers and note that nine divisions on the main scale are covered by ten divisions on the vernier scale, as shown in Figure 10–2A. The first vernier mark, indicated by **A** in the figure, is short of the millimeter mark by 1/10 or 0.1 mm. If the caliper were opened by 0.1 mm., this vernier mark would line up with the millimeter mark above it. When the instrument is closed, the second vernier mark is 0.2 mm. away from a millimeter line, and if the zero end of the vernier is 0.2 mm. past the millimeter mark, the second vernier line will line up with the one above it. The tenths of a millimeter are obtained by finding which vernier line coincides best with a line on the main scale. Figure 10–2B shows a vernier reading 5.6 mm. The 5 is read at the zero on the vernier and the sixth line coincides with a millimeter mark.

There are differences on different instruments as to the number of lines on the vernier. If 50 vernier divisions cover 49 main scale divisions, the vernier divisions read to 1/50 or 0.02 of a scale division. If the scale division corresponds to 0.5 mm., then a vernier with 50 divisions would read directly to 0.01 mm. In such an instance, the zero end of the vernier would read to the half millimeter. If the zero end of the vernier was past a millimeter mark but not as far as the half millimeter mark, the vernier would read directly. If the zero on the vernier was beyond a half millimeter mark, 0.5 would have to be added to the vernier reading.

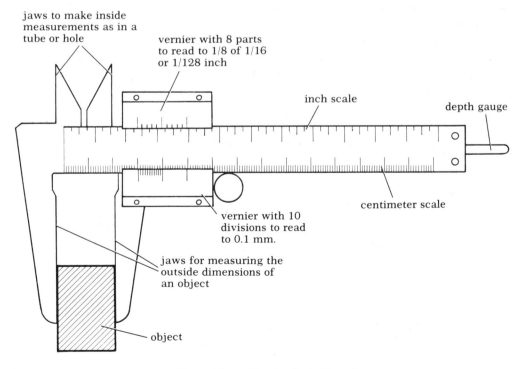

Figure 10–1 The vernier caliper.

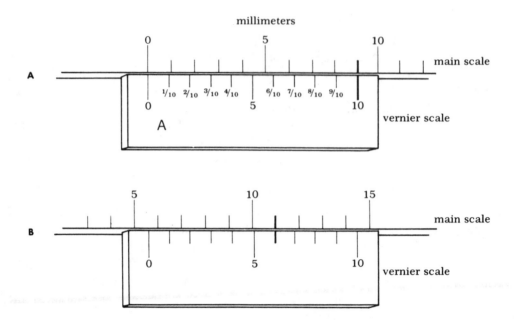

Figure 10-2 The use of a vernier scale. In **A** the jaws of the caliper are closed and it reads 0.0. In **B** the position of the vernier scale shows a reading of 5.6.

THE MICROMETER CALIPER

Another instrument for measuring small distances is the micrometer or screw caliper which is shown in Figure 10–3. The principle of the micrometer is that as a screw is turned by one revolution, it advances a distance equal to the pitch of the screw. A fraction of a rotation advances the screw by a corresponding fraction of the pitch. It is common in a micrometer to have a screw with a pitch of 0.5 mm. and a thimble to read to fiftieths of revolutions or 0.01 mm. Markings on a sleeve give the readings of the distance to 0.5 mm., and the reading on the thimble gives the number of hundredths to be added to the reading on the sleeve. Figure 10–4 shows two examples.

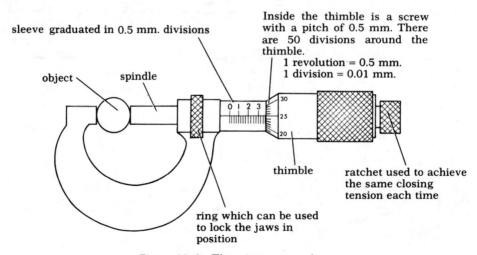

Figure 10-3 The micrometer caliper.

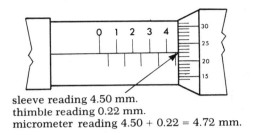

sleeve reading 4.50 mm.
thimble reading 0.22 mm.
micrometer reading 4.50 + 0.22 = 4.72 mm.

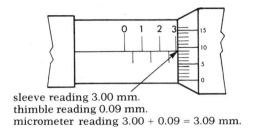

sleeve reading 3.00 mm.
thimble reading 0.09 mm.
micrometer reading 3.00 + 0.09 = 3.09 mm.

Figure 10-4 Two micrometer readings illustrated.

Micrometer calipers often do not read zero when the jaws are closed. To correct for this, a zero reading must always be obtained and subtracted from the micrometer reading of an object. If the zero reading is below the zero mark, it is called negative and is added. Examples of zero readings are shown in Figure 10-5.

A micrometer caliper should always be closed gently, using the ratchet if there is one. The screw has a high mechanical advantage and if the drum is forced, the object being measured may be squeezed by several hundredths of a millimeter or the jaws may be bent apart by a similar amount. Never force a micrometer.

THE TRAVELING OR MEASURING MICROSCOPE

To measure an object with a traveling or measuring microscope, one end of the object to be measured is lined up on the cross hairs of the instrument and the position of the microscope on a scale is recorded. The microscope is then moved to line up on the other end of the object and again the microscope position is recorded. It is important to always approach a microscope setting from the same direction in order to avoid errors as a result of what is called *backlash,* which results from a small amount of play in a screw. If you set the microscope on a mark by turning the screw in one direction and then watch as you turn it back, you will see that the screw turns a small amount before the microscope begins to move. As a result of this phenomenon, the scale reading depends on the direction the screw was turned to make each microscope setting. If you move the screw too far in making a setting, you must move the microscope well back so that the final setting is made in moving the screw in the same direction as for all the other readings. Choose a direction for the settings and stick to it! The distance measured is the difference between the two scale readings. This microscope has many advantages, such as being able to work at a distance from the object. One type of measuring microscope is illustrated in Figure 10-6. This type of microscope has

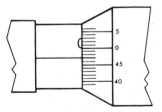

micrometer reading minus 0.03 mm.
to be added to each reading

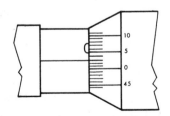

micrometer reading 0.03 mm.
to be subtracted from each reading

Figure 10-5 Two micrometer zero readings illustrated.

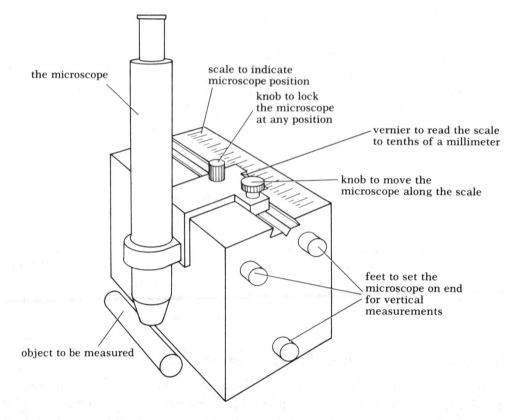

the microscope

scale to indicate
microscope position

knob to lock
the microscope
at any position

vernier to read the scale
to tenths of a millimeter

knob to move the
microscope along the scale

feet to set the
microscope on end
for vertical
measurements

object to be measured

Figure 10-6 A measuring microscope. To measure an object, move the microscope so that one end of the object is seen on the cross hairs and read the position on the vernier. Then move the microscope to focus on the other end of the object. The difference in microscope position readings give the length of the object.

only one scale, but for versatility the telescope can be mounted in three directions.

Another type of measuring or traveling microscope is illustrated in Figure 5-15. This model has two scales, one vertical and one horizontal. Each scale has a vernier reading to 0.01 mm. with the aid of an attached lens. This is one of the best instruments to use for the experiments outlined.

An eyepiece scale or graticule may be used to measure objects or movements small enough to be included in one field of view. In this case a scale is put into the microscope tube so that upon looking into it the viewer sees both the scale and the object. The size of the object can be expressed in these scale divisons. By using a scale of known length as an object, the viewer can calibrate the eyepiece scale. Scales on glass with millimeter divisions divided into tenths of millimeters and one tenth divided into hundredths are available for such calibrations.

In order to use a microscope properly, you should understand its construction. Figure 10-7 shows a cross section of a microscope with the functions of the principal parts labeled. At the end of this chapter an experiment is described in which a microscope is set up using only two simple lenses. Methods to measure magnifying power are also described and you will have an opportunity to use these methods.

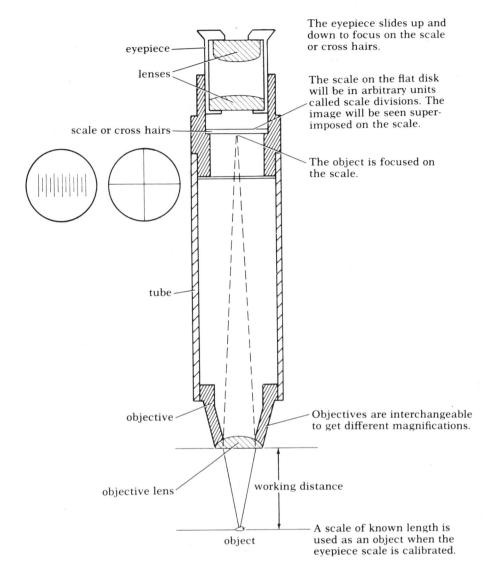

The eyepiece slides up and down to focus on the scale or cross hairs.

The scale on the flat disk will be in arbitrary units called scale divisions. The image will be seen super-imposed on the scale.

The object is focused on the scale.

Objectives are interchangeable to get different magnifications.

A scale of known length is used as an object when the eyepiece scale is calibrated.

eyepiece

lenses

scale or cross hairs

tube

objective

objective lens

working distance

object

Figure 10-7   A cross section of a microscope tube.

The proper focusing of a microscope is essential for its effective use because an improperly focused microscope will give inaccurate readings. The procedure outlined here does not include the focusing and alignment of substage condensers, since these are not used with a traveling microscope.

The eyepiece or ocular must first be focused on the cross hairs or eyepiece scale in such a way that the image seen is about 25 cm. (10 inches) from the eye, or at the distance for comfortable reading. This is accomplished by moving the eyepiece until the cross hairs are in clear focus and then making finer adjustments in one of several ways. One way is to look at an object on the table and then look quickly into the microscope tube. The cross hairs or scale should be immediately in focus. The eye muscles should not have to change the focusing of the eye as you look from the table into the instrument.

The more precise method is the method of parallax, which can be illustrated in this way: position two objects at different distances and apparently in line with the point from which you are looking at them. Move your head, and look at the objects from a different angle; they are no longer in line. As you move your head back and forth the objects seem to move back and forth relative to each other. This apparent relative motion of the object resulting from your (the observer's) motion is what is called *parallax*. If the two objects are then moved to the same distance, there is no motion between them, even when you move your head; parallax has been eliminated. Look into the microscope with one eye and at a book or object on the table with the other. Adjust the eyepiece until there is no parallax between the images of the book and the cross hairs. After the eyepiece has been set in this way, it should not be touched again.

To focus properly on an object, the image in the tube must coincide with the plane of the cross hairs. This is accomplished by using the focusing adjustment, which changes the object distance, until there is no parallax between the image of the object and the image of the scale or cross hairs. This has been achieved when, as you wiggle your head while looking into the microscope, the scale remains fixed on the object and the scale and object move together. If this is not achieved, the reading or setting will vary with the angle of viewing into the instrument, leading to inaccurate results.

THE CAMERA

The camera may be used to measure lengths, although, because of film and lens distortions, the accuracy is limited to a few per cent with an ordinary camera. The camera is such a valuable tool in science that a description of its use is worthwhile.

In using a camera there are two important procedures: one is obtaining the picture, which involves proper focusing and exposure, and the other involves the conversion of distances measured on the film to distances on the object. If measurements are comparative, that is, either compared with each other or with a scale also imaged on the film, there is no problem. However, this is not always the case, and therefore, some techniques to obtain measurements of the object in terms of measurement on the photograph will be discussed.

For a camera set to take pictures of distant objects the film is at the principal focus of the lens; the lens-to-film distance is the focal length, and this is usually marked on the lens mount. If measurements are to be made of objects that are distant enough so that the lens-to-film distance is just the focal length, the image size I is related to the object size O by $I/O = f/p$, where p is the distance from the lens to the object and f is the focal length of the camera lens. This is shown in Figure 10–8A.

For nearby objects the lens is moved forward to transfer the image onto the film. This forward lens motion on cameras will frequently allow focusing to about three feet. For closer objects one of two systems may be used. Either tubes or bellows may be inserted to move the lens even further out, or an auxiliary lens may be used. In the latter case, the camera may be set for infinity. When the camera is set for infinity, parallel rays entering the lens are focused onto the film. If an auxiliary lens is put in front of the camera lens and the object is placed at the focal distance of that lens, then light diverging from the object is formed into a

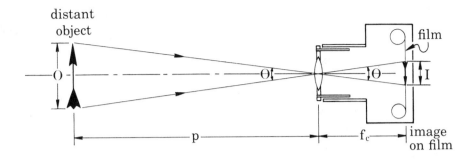

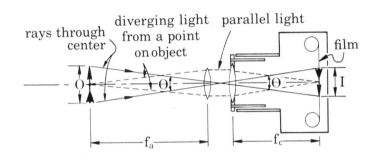

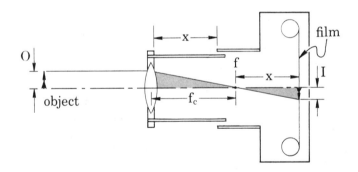

Figure 10-8 The use of a camera to find object sizes in terms of the image size on the film. For a distant object as in **A**, the relation is seen from the triangles formed by the rays through the center of the lens to be $O/I = p/f_c$. Using the close-up or auxiliary lens as in **B**, with the object at the focal point of the auxiliary lens, the relation is close to $O/I = f_a/f_c$. If the camera lens is moved forward a distance x to photograph a nearby object as in **C**, then by the shaded triangles it is apparent that $O/I = f_c/x$.

parallel beam to enter the camera and to be focused onto the film. The arrangement is as in Figure 10-8B. The ratio of the size of the image on the film to object size is very close to the ratio of the focal length of the camera lens to the focal length of the auxiliary lens. If the camera is not set to be focused for parallel rays, the calculation of image to object size is more complex and involves the lens equations. It is also necessary that the auxiliary lens be close to the camera lens.

If to obtain a photograph of a nearby object the camera lens is moved forward a distance x, then by reference to Figure 10–8C it is seen that the ratio of image size to object size is given by x/f_c. This relation applies even if the lens is thick. In the case of a thick lens, the position from which to measure image and object distance is not clear, but the value x is easily and precisely measurable. The focal length of the camera lens is a precise quantity marked on the lens mount. For close-up photography the range of sharp focus is usually so short that the above relation allows measurement of size to within 2 or 3 per cent.

To obtain a proper exposure under ordinary conditions, the manufacturer's recommendations, which come with the film, or an exposure meter can be referred to. The variables are the exposure time, the size of the aperture at the lens and the sensitivity of the film.

The available shutter speeds on a camera, that is, the lengths of time that the film is exposed to light, are usually such that there is a factor of two between adjacent ones.

The diameter of the lens opening is indicated by what is called the f number or the focal ratio. The diameter, d, is expressed as a fraction of the focal length. For example, a setting of $f/8$ indicates that the diameter of the opening is one-eighth of the focal length. The intensity of the light on the film depends on the area of the opening, which is directly proportional to d^2, and inversely proportional to the square of the distance to the film. This will be an image distance q or $x + f$ as shown in Figure 10–8. Except in close-up work, q will be very close to f. The intensity on the film, then, is proportional to d^2/f^2. Since $d = f/N$, where N is the focal ratio, the light intensity is inversely proportional to the square of N. The light intensity at $f/8$ would be twice as much at $f/11$, since 8^2 is just about half of 11^2. The focal ratio settings on a camera are usually such that in going from one to the next, the change in intensity is a factor of two.

The total amount of light reaching the film during an exposure depends on the ratio T/N^2, where T is the exposure time. The same value of T/N^2 will yield the same final result, except during long exposures for which the film characteristics cause a change, or for close-up pictures for which the lens-to-film distance is not equal to f. If the distance is q, the exposure must be increased by the ratio q^2/f^2.

Another system of indicating exposure is the "Exposure Value" or EV system. If the ratio T/N^2, with T in seconds, is calculated and expressed in the form

$$\frac{T}{N^2} = \frac{1}{2^n} = \frac{1}{2^{EV}}$$

The value of the exponent of the 2 is the EV number. For each increase of one unit in the EV number, the exposure is decreased by a factor of two.

MEASURING WITH LIGHT WAVES

Measurement using light waves is one of the more elegant procedures of physics. The wavelengths of light are in the range from about 0.00006 cm. for red to about 0.00004 cm. for violet, so a wave amounts to a very small unit of length. Not only that, but some elements produce a spectrum having only certain wavelengths, each of which is very precise. The standard meter is defined, in fact, as

1,650,763.73 wavelengths in a vacuum of a certain orange-red line in the spectrum of the isotope krypton 86. On October 14, 1960, when this definition of the standard meter was adopted, the platinum-iridium bar which had been a standard since 1889 was discarded as the fundamental standard, although it is still of practical use and is referred to as the "International Prototype" meter.

The procedure for using light waves to measure physical objects consists of two parts: first, the determination of the wavelength of the light being used, and then the devising of a means to use the waves for measurement.

There are two basic ways to determine wavelengths. One way is to compare the unknown wavelength to a known length, and the other way is to compare the unknown wavelength to a known wavelength. In the first instance, the obvious solution is to count the number of waves in one meter. This is effectively what was done to obtain the new definition of the standard meter. The number of waves in the distance between the two marks on the standard meter bar was precisely determined. One of the reasons for the change in the standard was that if by some accident the meter bar were to be damaged or destroyed, the international standard meter still would not be lost. Furthermore, the fundamental standard of length is now available in any laboratory that can set up the necessary apparatus.

Unfortunately, light waves can not be made to align themselves along a meter bar and remain still long enough to be counted like so many pennies. The techniques used are not so direct. Many methods to relate wavelength to distance have been devised and a few of them will be described.

All of the techniques make use of the property known as interference. When two waves of equal amplitude combine, the result is a wave motion with an intensity anywhere from zero to four times the intensity of one of the waves alone (why four, not two?). The intensity of the combination depends upon whether the waves are "in step" or "out of step," the correct terminology being *in phase* or *out of phase*. In general, then, the intensity depends on the phase difference between the waves. The idea used is that a beam of light is separated into two parts which travel different paths and then are recombined. If the paths are exactly the same length, the waves will be in phase when they recombine, but if there is a difference in the path length, the phase difference will depend on the path difference. Sometimes when the reflection occurs, the wave is given a sudden change in phase equivalent to an extra path of a length equal to half a wave. This usually occurs when the wave is going from a medium of low index of refraction to a medium of high index. This phase change is taken into account by adding half a wavelength to the geometrical path difference.

One of the simplest devices for comparing light waves with a distance consists of only two flat glass plates in smooth contact along one edge and separated at the opposite edge by the object, the thickness of which is the distance for comparison. The arrangement, as shown in Figure 10-9, is placed on a black surface and is illuminated and also viewed perpendicular to the surface. To achieve this the light is reflected off an oblique glass surface onto the plates which produce the interference of the waves. The plates are then viewed through this reflecting plate. Monochromatic light (light of one wavelength) must be used. The glass plates must have surfaces flatter than that of ordinary window glass; commercial plate glass is satisfactory.

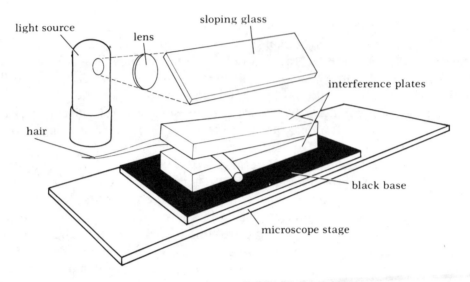

Figure 10-9 The use of two flat glass plates (interference plates) to compare the size of an object to the wavelength of light. The object separates the plates at one end.

Of the light that strikes the glass plates, some is reflected from the top surface of the thin air wedge between the plates, and some is reflected from the bottom surface where the light enters the lower plate. At this second reflection there is a change in phase. The light emerging from the glass plates is the sum of that reflected from the two surfaces. In some places the light will be in phase and in other places it will be out of phase, depending upon the path difference. At a position where the thickness of the air wedge is t (see Figure 10–10) the geometrical path difference is $2t$ and to take into account the change of phase at one of the surfaces, the path difference is effectively $2t + \lambda/2$, where λ is the wavelength. Constructive interference (bright light) will be seen if the path difference (p.d.) is equal to $n\lambda$, where n is an integer or zero. Destructive interference (no light) will be seen at the positions where the path difference is equal to $(n + 1/2)\lambda$. So the fringes are bright where

$$2t + \lambda/2 = n\lambda$$

or

$$t = n\lambda/2 - \lambda/4$$

It is apparent from this that in going from $n + 1$, t increased by $\lambda/2$, or half a wavelength. The observed effect is that the reflected light shows a series of light and dark bands, or fringes, as seen in Figure 10–11. In the distance from one fringe to another, the air wedge changes in thickness by half a wavelength. If there are N bright fringes between the end of the glass plates where they touch and the position of the space, then the thickness, T, of the spacer (the object) is given by

$$T = N\lambda/2 - \lambda/4$$

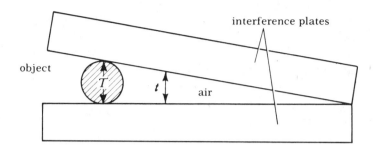

Figure 10-10 The air wedge between the two plates used to measure an object in terms of light waves.

N can be either counted or found from the number of fringes per unit length. If T is measured, say, with a micrometer caliper, then λ can be determined; or if λ is known, T can be determined.

The most common laboratory instrument for the measurement of the wavelengths of light is the diffraction grating spectrometer. Diffraction gratings may be for reflection or transmission of light. Reflection gratings consist usually of a piece of polished metal on which is ruled a large number of closely spaced parallel lines. The transmission grating, which will be considered here, is similar but is ruled on a transparent material such as glass. The spaces between the lines act as very narrow slits through which the light passes. Light, being a wave motion, spreads out when it goes through a slit — the phenomenon being especially noticeable for very narrow slits. This phenemenon, illustrated for water waves in Figure 2-1, is called diffraction. If light is put normally through N slits of a grating and observed at an angle from the direction of the incident light, then N separate beams of light are being observed and there are path differences between all of them. There will be a particular direction at which the path difference from each

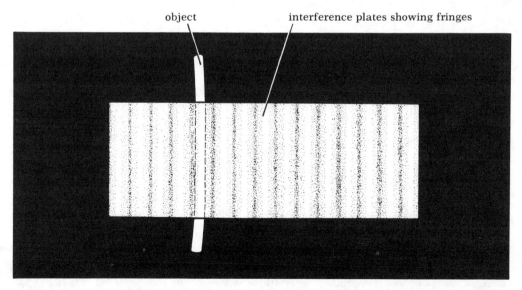

Figure 10-11 The light and dark banks, or fringes, seen with the apparatus shown in Figure 10-9.

adjacent slit is one wavelength, λ, and in this direction the light will be bright because that coming from all the slits is in phase. (See Figure 10–12A.) In another direction, in which the path difference for the light from adjacent slits is 2λ, the intensity will again be a maximum (Figure 10–12B). In general, if the path difference is n, bright light will be observed. These directions in which the light is bright depend on the wavelength, λ, so different wavelengths or colors are bright in different directions. The diffraction grating separates all the wavelengths or colors in the incident light and they go in different directions. This is similar to the action of a prism except that a number of spectra may be produced on each side of the incident beam.

If the spacing between the slits is s, then from an analysis of Figure 10–12C it follows that for the nth spectrum or "order," $n\lambda/s = \sin \theta_n$ or $n\lambda = s \sin \theta_n$.

If s is known and θ is measured (n is readily determined), the wavelength can be calculated. The slit spacing s is merely the reciprocal of the number of lines per unit length across the grating. Gratings of 15,000 or even 30,000 lines per inch are common. It goes without elaboration that the ruling of this number of straight, parallel, evenly spaced lines per inch (and some gratings have ruled surfaces 6 inches across) is one of the most delicate operations yet undertaken.

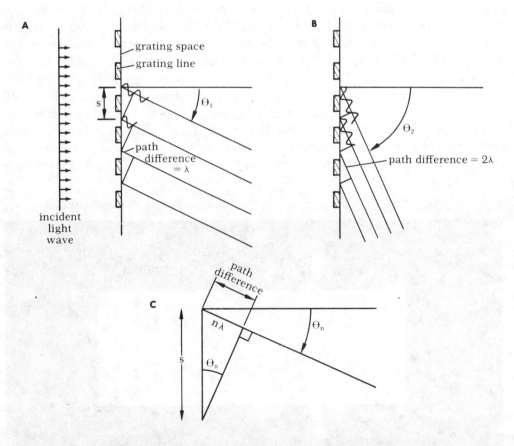

Figure 10-12 Diagrams to illustrate the analysis of the diffraction grating. An end view of the lines is shown, greatly magnified. Bright light is seen in the directions for which the path difference from adjacent slits is one wavelength as in **A**, two wavelengths as in **B**, or n wavelengths where n is a whole number. From **C** you see that $n\lambda = s \sin \theta_n$.

A grating may be used to compare an unknown wavelength to one that is known. Using the formula developed above, we can easily show that in any given order, an unknown wavelength λ_2 compares to a known λ_1 in the manner.

$$\lambda_1/\lambda_2 = \sin \theta_2 /\sin \theta_1$$

In a given spectrum if one wavelength is known, the others may be found.

There have been thick books devoted to tables of accurately determined wavelengths, and yet there is still work being done in this area. Two well-established wavelengths used in introductory laboratories are the bright green line in the mercury spectrum, 546.1 mμ and the sodium yellow line, 589.3 mμ. This sodium line is actually a doublet with components at 589.0 and 589.6 mμ. In many experiments the fact that it is a doublet can be disregarded, but others it cannot be.

The measurement or comparison of wavelengths with a grating involves the measurement of angles, which will now be described.

MEASUREMENT OF ANGLES

The protractor is the familiar instrument for the measurement of angles, but the ordinary protractor is limited to measurement to the nearest degree or, at best, to half of a degree. In the study of spectra, for instance, it is necessary to measure angles far more precisely than this, and some spectrometers are fitted with a protractor having a vernier to read fractions of degrees. Different instruments will have different precision. The spectrometer used by an introductory student may read to tenths of a degree, the main protractor being marked in degrees and the vernier having ten divisions. If the vernier caliper has been mastered, there will be no difficulty with the reading of such a spectrometer. More precise instruments may read to the nearest minute of angle. This is often accomplished with a main scale marked to half degrees and the vernier having 30 divisions. It is necessary with any particular instrument to study the main scale and the vernier in order to determine how to make readings.

There is, of course, more to the spectrometer than just the scale. The basic construction of a spectrometer is shown in Figure 10–13. The light source is viewed through the telescope. If a prism or diffraction grating is placed on the table, the direction of the light coming through the collimator from the source will be changed. The telescope can then be moved to the appropriate direction, as shown before in Figure 8–10, and the angle by which the light is deviated can be determined. To do this, the vernier is made to move with the telescope and its position on the protractor scale can be determined. The light striking the prism (or diffraction grating, as the case may be) must be parallel so that slit A of the collimator must be at the focal distance from lens B. The telescope must also be focused for parallel light so that light diverging from the slit is focused onto the cross hairs, D, of the telescope. The eyepiece magnifies the image of both the slit and the cross hairs.

The adjustment of the spectroscope is in two parts: the adjustment of the telescope and then the collimator.

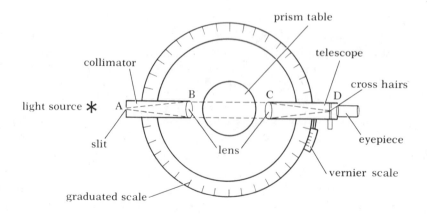

Figure 10-13 The basic parts of the spectrometer.

To adjust the telescope, look into the eyepiece and move it in and out until the cross hairs are clear and their image is at a comfortable distance from the eye. Point the telescope at a distant object and adjust the objective lens until the image of the distant object is in the plane of the cross hairs. Then both the cross hairs and the object will be in focus at once and there will be no relative motion as the eye is moved; that is, there will be no parallax. The telescope is now adjusted for parallel light. Do not change the adjustment of the objective lens. Only the eyepiece may be adjusted to suit the individual.

To adjust the collimator place a light source in front of the slit and look toward the collimator with the adjusted telescope. Adjust the collimator lens until the image of the slit is in the plane of the cross hairs. This is accomplished when the slit and the cross hairs are simultaneously in focus for your eye and when there is no relative motion (parallax) between them when your eye is moved. The instrument will now be in adjustment. The light from the collimator will be in a parallel beam and the focusing of the eyepiece is such that it will not be tiring to look into it for long periods. Parallax will be eliminated, so the settings will not depend upon how you look into the instrument.

Other adjustments are described in Figure 10-14. Except for the knobs to move the telescope (*B, C*) or prism table (*E*), or the knob to vary the slit width (*G*), the student should not have to touch these adjustments. If the slit cannot be properly focused on the cross hairs, help should be obtained from the laboratory instructor.

One very important point about such an instrument is that movable parts should move with little effort. Never force the telescope to move. If it is tight, the cause is probably that it has been locked by a knob. Find the knob and loosen it. Forcing the telescope will damage the instrument.

ANGLES MEASURED WITH A LINEAR SCALE

It is frequently more precise to measure an angle in an experimental situation by finding the lengths of two sides of a triangle and then using the appropriate trigonometric function. But there is another system for angular measurement, which is also frequently used in scientific work. This involves the *radian*, which is

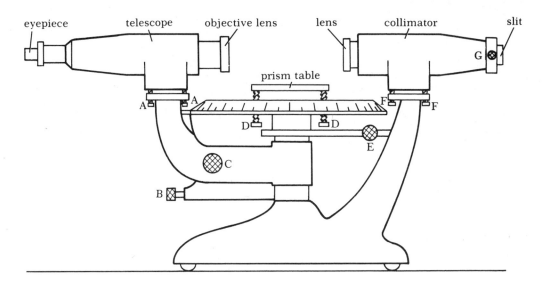

Figure 10-14 A sketch of a spectrometer showing the various adjustments. *A*, Screws to level or align the telescope. *B*, Knob to fix the position of the telescope. This knob must always be loosened before moving the telescope by hand and locked only to make fine adjustments with knob C. *C*, Fine adjustment knob. *D*, Screws to level or raise the prism table. *E*, Knob to fix the position of the graduated disc and prism table. When readings are taken with the telescope, this knob must be in the locked position. *F*, Screws to align the collimator. *G*, Knob for adjusting the slit width. The slit should be wide enough to allow a sufficient amount of light to pass through but as narrow as possible to allow accurate measurement.

arrived at as follows: referring to Figure 10–15A, the angle θ is to be measured, and s_1 and s_2 are arcs drawn with radii r_1 and r_2. The arc length for the larger radius is proportionately larger, and the ratio of arc length to radius does not depend on the radius chosen. That is, the ratio s_1/r_1 is the same as the ratio s_2/r_2. This ratio is merely a function of the angle and can be used as a measure of the size of the angle. In Figure 10–15B a larger angle is shown. The ratio s/r for this angle will be proportionately greater than the ratio for the smaller angle in Figure 10–15A. *The angle in radians is defined as the ratio of arc length to radius.* An angle of one radian is quite large; it is the angle for which the arc length is equal to the radius.

The conversion factor between degrees and radians can be found from the angle in a complete circle. This is 360°, but how many radians? Referring to Figure 10–15C, it can be seen that for a circle the arc length is $2\pi r$, and the angle in radians is given by

$$\theta = \frac{2\pi r}{r} = 2\pi$$

That is, there are 2π radians in a circle. Also, there are 360° in a circle, so

$$2\pi \text{ radians} = 360°$$

or
$$1 \text{ radian} = 360°/2\pi = 57.3°$$

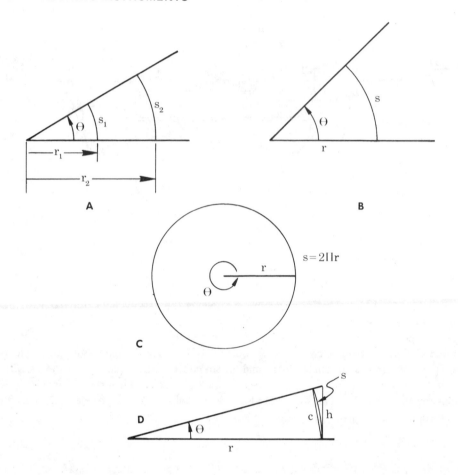

Figure 10-15 Radian measure of angles. In A, $s_1/r_1 = s_2/r_2$ and is a measure of θ in radians. In **B**, s/r is greater than in **A** and so is the angle. The angle in a circle shown in **C** is $2\pi r/r = 2\pi$ radians. For a small angle as in **D**, the ratios s/r, c/r and h/r are almost identical.

also
$$1° = 1/57.3 \text{ radians} = 0.01745 \text{ radians},$$

$$1' = 1/60 \text{ degree} = 0.000291 \text{ radians}$$

and
$$1'' = 1/60' = 0.00000485 \text{ radians}$$

The principal use of radian measure occurs with small angles. With a small angle, preferably smaller than that shown in Figure 10-15D, the curved length s is very nearly the same as the chord c or the perpendicular distance h. The ratios s/d, c/d, and h/d are all very similar. To show how small the differences really are, these ratios are shown for some small angles in Table 10-1. Even at $10°$ the difference is only half a percent. To show how this is used, if in a particular situation s is 1.0 mm and r is 1000 mm, then θ is 0.0010 radians or 3.3 minutes of angle.

TABLE 10-1 A Comparison of the Quantities θ in Radians, Sin θ, and the Ratio, Chord/Radius.

θ Degrees	$\theta = \dfrac{s}{r}$ Radians	$\dfrac{c}{r} = \dfrac{\text{Chord}}{\text{Radius}}$		$\dfrac{h}{r} = \sin\theta$	
		Value	Error	Value	Error
1°	0.0174533	0.0174531	−0.0001%	0.017452	−0.002%
2°	0.034907	0.034905	−0.005%	0.034899	−0.023%
3°	0.052360	0.052354	−0.01%	0.052336	−0.046%
4°	0.069813	0.069799	−0.02%	0.069756	−0.08%
5°	0.087266	0.087239	−0.03%	0.087156	−0.1%
10°	0.174533	0.174311	−0.13%	0.173648	−0.5%

ELECTRICAL MEASUREMENTS

MOVING-COIL METERS

The principle employed in one of the most common types of voltmeter or current meter is that there is a force exerted on a current-carrying wire in a magnetic field. The heart of the meter will often be a movable coil in a magnetic field. The current to be measured is put through this coil. Attached to the coil is the pointer which moves across the scale as the coil turns until the torque due to the force on the wire is balanced by a fine helical spring. The coil itself is usually mounted on jewel bearings. Figure 10–16 shows the movement of this D'Arsonval type of instrument.

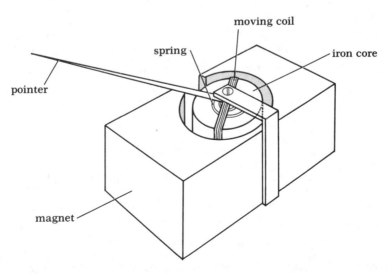

Figure 10-16   The movement of a D'Arsonval type of meter.

The force exerted on the sides of the coil when a current is passed through it depends on the length of the coil in the field, the number of turns, the strength of the magnetic field, and the current in the coil. The torque depends on the force and on the force arm, which is half the width of the coil. The angle through which the pointer moves depends on the stiffness of the spring. All of these things, except the current, are constant for a particular meter, so the deflection depends on the current through the meter. The basic part of the moving-coil meter is thus a current-indicating device, whether the face of the meter indicates current or voltage.

A movement of high sensitivity can be converted by the addition of external resistances to a meter reading a wide range of current or voltage. Consider, for example, a meter movement which will indicate full scale deflection when 50 microamps (1 microamp = 1 μA = 10^{-6} amp) pass through the coil. In order to attain such sensitivity, many turns of very fine wire will be used for the coil and the resistance may be about 1000 ohms. If you desired to measure a current up to 500 μA, the meter would be of no use as it stood: It would be too sensitive. However, a resistance could be placed across the meter terminals and could be of such a size that when 500 μA flowed toward the meter, 450 μA would be shunted through the added resistance and 50 μA would flow through the meter movement. So with the shunt resistance in place, the meter scale would indicate from 0 to 500 μA flowing in the circuit towards the meter.

The value of the shunt resistance is easily chosen. In Figure 10–17A I is the current to be measured, I_m is the current in the meter movement, and I_s is the current in the shunt. In our example, we want I_m to be 0.1 I. Then $I_s = 0.9/I$ and $I_s/I_m = 9$. Because current varies inversely as resistance, the resistances R_s and R_m are in the ratio of $R_m/R_s = 9$, or $R_s = R_m/9$. That is, if the meter resistance is 1000 ohms, a shunt resistance of 1000/9, or 111 ohms, would be used to make that 50 μA meter movement give full scale deflection for 500 μA.

If the meter is to read up to 5000 μA, or 5 mA, the ratio of currents would be $I_s/I_m = 99$ and $R_s/R_m = 1/99$. If R_m is 1000 ohms, R_s would be 11.1 ohms. By the proper choice of shunt resistances, the meter can be made to indicate full scale for any current higher than that required to make the basic meter movement show full scale deflection.

The movement described can also be made to indicate voltage. If a meter which indicates up to 100 volts is desired, a resistance can be put in series with the meter, as shown in Figure 10–17B. The resistance should be of such a size

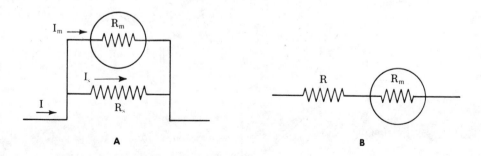

A   B

Figure 10–17 The adaption of a sensitive current meter to have various current ranges using shunt resistors as in **A**, or to have various voltage ranges using series resistors as in **B**.

that with 100 volts across the system a current of 50 μA will flow. Then using $R = V/I$, when V = 100 volts, I will be 50 μA or 50 $\times$ 10^{-6} amp and R will be 2,000,000 ohms. To make the total resistance equal to 2,000,000 ohms, 1,999,000 ohms would have to be added to the 1000 ohm meter resistance. It is left as an exercise for you to show that for a full deflection of 1 volt the total resistance would have to be 20,000 ohms, made up of 19,000 ohms added to the 1000 ohm meter resistance. For full scale deflection by 2 volts, the resistance would have to be 40,000 ohms. Such a meter would be rated at 20,000 ohms per volt. For each volt required for full scale deflection, the resistance would have to be 20,000 ohms, and this is called the sensitivity of the meter. The higher the sensitivity, the higher the number of ohms per volt.

A variety of shunt or series resistances can often be chosen for one meter movement by means of a switching arrangement. Such a meter is a very versatile instrument and because of the large number of possible ranges, it is called a multimeter. Figure 10–18 is a reproduction of the circuit for the Starkit multimeter MT 6-C. This meter is rated at 20,000 ohms per volt for direct voltage. Trace the circuit for the switch in the 120 volt position and verify this.

One general characteristic is that current meters have low resistance and the voltage across them is small, whereas voltmeters have very high resistance and draw little current.

The meter shown in Figure 10–18 has switch positions for the indication of alternating voltage and also resistance. The meter cannot indicate alternating current or voltage directly because the direction of the force on the coil would change as frequently as the alternations (120 times per second for 60 cycle voltage). The needle would not have time to move. To get around this difficulty a rectifier is often used, either half wave as in Figure 10–19A of full wave as in Figure 10–19B. These rectifiers are often small crystal rectifiers inside the meter.

An A.C. meter which uses a rectifier produces a deflection which is proportional to average current or voltage. What is more important in practical applications, however, is the root mean square, or r.m.s. value. The markings on the dial of a rectifier type of meter can be placed to indicate r.m.s. values for one particular wave form. Invariably the sine wave is chosen, and the voltage reading would not be correct for any other wave form. Fortunately, the sinusoidal variation is most common and the use of the rectifier type of meter is justified. Another type of meter does have a deflection proportional to the square of the current, and the numbers on the dial are marked as the r.m.s. values. In this type of meter the magnetic field is produced by the current in a stationary coil and the current is also put through a moving coil as in the D.C. D'Arsonval type of meter. The torque is proportional to the current in the coil and to the magnetic field, which is also proportional to the current. The net result is that the torque depends on the square of the current. The scale on this type of meter is compressed toward the zero end and expanded at the upper end.

The moving-coil meter may also be adapted to measure resistance. To accomplish this, batteries are inserted as shown in Figure 10–18. The resistance to be measured is connected across the meter terminals and the amount of current which flows depends on the resistance. Such a meter will have a special scale graduated in ohms. The scale may be made to cover various resistance ranges by using a selection of resistances in the meter.

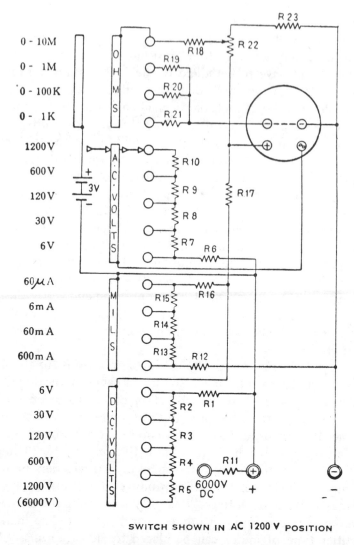

SWITCH SHOWN IN AC 1200 V POSITION

PARTS LIST

R 1	114.5K	R 9	4.8M	R17	3.1K
R 2	480K	R10	6M	R18	55K
R 3	1.8M	R11	96M	R19	6.7K
R 4	9.6M	R12	1.8 ohms	R20	600 ohms
R 5	12M	R13	16 ohms	R21	59 ohms
R 6	57.5K	R14	150 ohms	R22	5K(V.R.)
R 7	240K	R15	16.5K	R23	17K
R 8	900K	R16	3.2K		

(K = Kilo = 1,000; M = Meg = 1,000,000)

Figure 10-18 A schematic diagram for a commercial multimeter. (Courtesy of Stark Electronics, Ajax, Ontario, Canada.)

A meter movement similar to the moving-coil type is the moving-iron type. The current to be measured is passed through a stationary coil inside of which is a small piece of soft iron pivoted and held by a small helical spring at an angle of 60 degrees to 90 degrees from the axis of the coil. The pointer is attached to this iron. A current in the coil produces a magnetic field and the iron experiences a

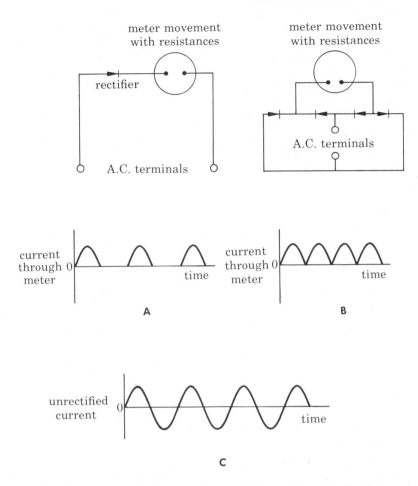

Figure 10-19 The use of rectifiers to adapt to D.C. current meter for the measurement of A.C. Diagram **A** shows half wave rectification, and **B** shows full wave rectification.

torque to align it with the field, resulting in a deflection of the needle. Such an instrument can be adapted as a voltage meter also, but the sensitivity of such a meter is, in general, not as great as that of the D'Arsonval movement.

Most multimeters, which will often measure many ranges of direct or alternating current and voltage (and frequently ohms also), are not instruments of high precision, but they are adequate and convenient for a multitude of laboratory work. The main advantages are versatility and, very often, high sensitivity. The main causes of low precision are variations in the internal resistances (frequently associated with misuse) and pivot friction, which is increased with wear and rough handling. Instruments designed to better overcome friction will do so partly by having increased torque, which is obtained by using higher current. Such an instrument would then have low sensitivity, so in many applications the meter would disturb the circuit. If the meter resistance is known the effect of the meter can be taken into consideration, for it is just another resistance in the circuit.

Your first look at a multimeter may be a little frightening, but learn to use it and you will lose your fear. There will be one or two knots to set for what you want to measure. These switch in the appropriate series or shunt resistance. Turn it to read **AC** volts, **DC** volts, **DC** current, or ohms, choosing the proper range for your needs. It is actually better to start on too high a range and work down, just for the protection of the meter. A little examination will show the scale you are to read.

Increased precision is often obtained by using an entirely different approach. In fact, this is often the way with any problem — instead of striving to improve one technique try another way entirely. Both approaches have their place. Such variations in approach are the bridge type circuits, such as the Wheatstone bridge and the potentiometer circuit. Each of these circuits is widely used in electrical measuring instruments. The Wheatstone bridge will be described in detail.

THE WHEATSTONE BRIDGE

The moving-pointer type of electrical measuring instrument, although very common, is of fairly low precision and in some instances it has undesirable characteristics. Unless a mirror is provided to eliminate parallax when reading the scale, even 1 per cent accuracy is not obtained, although special laboratory types of meters may have a precision of 0.2 per cent of full scale. One sometimes undesirable characteristic is that current is required to operate the meter and when one is attempting to read voltage, this can lead to incorrect readings. A class of instruments which do not rely to a similar extent on pointer deflection is represented by the Wheatstone bridge and the potentiometer. These will often use a sensitive current meter to indicate no current, but nevertheless they are capable of far greater precision than the usual moving-pointer types.

The bridge type circuit has many variations which are used in such areas as nuclear radiation, temperature measurement, and light measurement.

The Wheatstone bridge as encountered at the introductory level is not direct reading. It requires a calculation to obtain the final reading and in this way it differs from the accustomed type of measuring instrument. It may be compared to a voltmeter having a scale marked in degrees of angular deflection so that each degree reading would have to be multiplied by a certain factor, depending on the construction of the instrument, in order to find voltage. Actually, the bridge can be made to be direct reading and this is often done in commercial models.

The foregoing has been written assuming that the reader knows what this instrument is, but to review, the following description is presented.

The Wheatstone bridge is basically a device to compare an unknown resistance to a standard resistance. All measurements are comparisons to some type of standard. Even a measurement of a length is a comparison of the unknown length to a standard of length, and it is a mental exercise to think of how all measurements are really comparisons. The basic Wheatstone bridge is illustrated in Figure 10–20A. R_1 may be the unknown resistance and R_2 the standard. When the resistances are properly chosen so that no current flows in the galvanometer, voltage V_A at point A is the same as voltage V_B at point B. Then $i_1 = i_2$ and $i_3 = i_4$. The voltage $i_1 R_1 = i_3 R_3$ and $i_2 R_2 = i_4 R_4$. These relations are combined to give the common Wheatstone bridge equation: $R_1 = R_2 \times R_3/R_4$. The accuracy of the determination of R_1 depends on the accuracy of the standard, R_2, and of the ratio R_3/R_4. In variations of the Wheatstone bridge, either the ratio R_3/R_4 is fixed and R_2 is variable or R_2 is fixed and the ratio R_3/R_4 is variable. A calculation is then necessary to obtain R_1. If the ratio R_3/R_4 is 1, then the unknown,

R_1, is just equal to R_2, and if R_3/R_4 is 10, then R_1 is just $10 \times R_2$. This principle allows simple calculation, almost direct reading of R_1. If R_2 is a non-variable standard, then the ratio R_3/R_4 can be made variable to balance the bridge. One common way to achieve this is to use a long wire with a movable contact for R_3 and R_4. This leads to the slidewire type of bridge, as illustrated in Figure 10–20B. If the resistance per unit length of the wire is r then $R_3 = rl_1$, and $R_4 = rl_2$. This ratio, R_3/R_4, reduces to l_1/l_2, and the unknown R_1 is given by $R_1 = R_2 l_1/l_2$. The accuracy of measurement of R_1 then depends on the accuracy of the measurements of the lengths l_1 and l_2.

In the student type of bridge in which the wire is often 1 meter long, the balance position should be obtainable to ±0.5 mm. and then if the balance point is near the middle, l_1 and l_2 are found to ±0.1 per cent and the ratio to ±0.2 per cent. The standard, R_2 should be chosen for at least this accuracy so that even with a student type of bridge, the accuracy is as good as the best of moving-pointer type instruments.

Now consider the case in which such a bridge is balanced so that the galvanometer reads zero. Then the temperature of R_1 is changed. Resistance changes with temperature, so the bridge becomes unbalanced and the galvanometer shows a deflection. If the galvanometer used is very sensitive (capable of reading micro-amps), then small changes in R_1 may produce large readings on the instrument scale. The bridge used in this way can measure the resistance and changes in that resistance, which is similar to measuring changes in the length of a rod as it is heated. This unbalanced bridge idea is made use of in a multitude of instruments, and this example is adaptable to a resistance thermometer. The bridge can be balanced when R_1 is $0°$ and then temperature changes in R_1 unbalance the bridge and show as a current reading. V_o could even be varied until the microamp readings correspond numerically to the temperature of R_1. In this adaptation, R_2 must be kept at a constant temperature, or if the material of R_2 has a low coefficient of resistance change with temperature, then it may be kept just approximately constant, perhaps at room temperature.

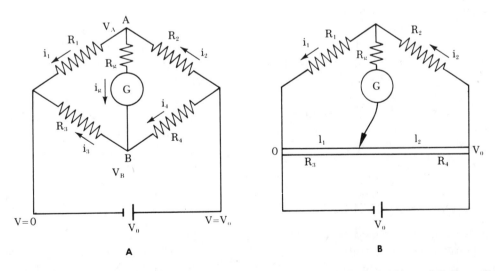

Figure 10–20 The Wheatstone bridge. Diagram **A** shows the basic bridge, and **B** shows the slidewire type of bridge.

MEASUREMENT OF RADIATION: LIGHT AND GAMMA RAYS

Today we speak of and use electric eyes, solar cells, ion chambers, Geiger counters, and so on. These all make use of what we call the photoelectric effect. Other effects are also used in the detection of gamma rays but they are not very different from the photoelectric effects.

Our knowledge of this photoelectric effect, which is the transference of the energy of a photon or "particle" of light to an electron which is bound to an atom, began about 1887 when Hertz, while experimenting with "radio" waves, noticed that one electric spark could trigger another. It was a long trail from the observation of this incidental phenomenon to the interpretation of it, and it took the insatiably inquisitive minds attributed to scientists to pursue such a trivial thing. The phenomenon was gradually realized to be due to ultraviolet light, although it worked with visible light on some materials like zinc. The same phenomenon was attributed to the discharging of an electroscope by light, either invisible ultraviolet light or visible light, depending on the surface.

The explanation had to await the discovery of the electron in 1897 and the postulation of light quantum by Planck in 1900. Finally, the process involved was suggested by Einstein in 1905. R. Millikan in 1916 did the crucial experiment proposed by Einstein to validate the theory. The intervening years had been a time of gathering data about the phenomenon — the effect of light of different wavelengths, the energies of the electrons ejected, the effect of varying the material, and so on — until a basis had been provided for a theory. Then when the phenomenon was understood, at least to some extent, the way was paved to make wide use of it.

Little things can be hints at important and useful processes. But who can predict? How many people over the years had seen the phenomenon but made no note of it? W. L. Lawrence, in an essay entitled "The Miracle that Saved the World,"[1] describes how this blindness to an unexpected phenomenon prevented the much earlier discovery of nuclear fission and the probable development of atomic weapons in Hitler's Germany.

Getting back to the photoelectric effect, it was noticed that the rate of ejection of electrons from a light sensitive surface depended on the intensity of the light. Here then was a method to measure, or at least to compare, light intensities — a method free of the judgment involved in the use of the human eye. There is now quite a variety of photocells and a few of them will be described. Unfortunately, because we cannot construct cells that react ideally, there is not usually a perfectly linear relation between light intensity and current flow, so the cells must be calibrated. A calibration procedure is described with each experiment that requires it.

TYPES OF PHOTOCELLS

The vacuum type photocell is the simplest, consisting of a photosensitive surface and a collecting electrode all enclosed in an evacuated glass envelope. One form is illustrated in Figure 10-21A pictorially and in Figure 10-21B schematically. The rod in the center is charged positively and is called the anode,

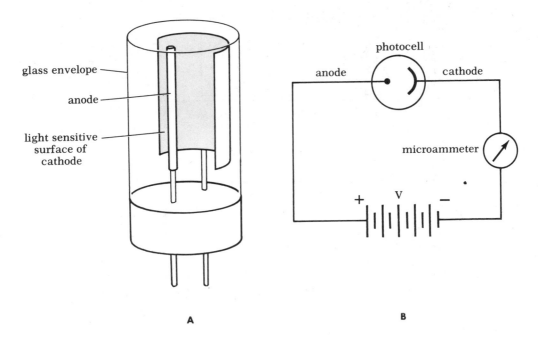

Figure 10-21   The vacuum type photocell. **A** is a drawing of one form of the cell and in **B** it is shown schematically in a diagram which also shows the voltage supply and the micro-ammeter to measure the current produced when light falls on the photocathode.

whereas the photosensitive surface (located on the inside of the semicylinder and coated with potassium or cesium) is negatively charged. The ejected electrons travel through the vacuum to the anode and then through the circuit to produce a current in the meter. The current from such a cell is usually in a range up to a hundred microamps. The collecting voltage may vary from 30 to 100 volts with little change in photocell current because only the electrons ejected by the light are collected and measured.

The gas-filled type of cell is a variation of the vacuum type. As implied, it is filled with gas, but the gas is of low pressure and of a type that will not react with the photosensitive surface. The photoelectrons are attracted and thereby acceler-ated toward the anode. They don't go far before they hit an atom of gas with sufficient energy to knock an electron from it. Then there are two electrons. Each of these is accelerated and each will hit an atom of the gas, knocking out elec-trons. Then there will be four. This is repeated, and it is common to have a tube in which an average of ten electrons are collected for each photoelectron emitted. This process is called *gas amplification.* Whereas such a cell is more sensitive than a vacuum type, the amplification depends on the applied voltage. It is, therefore, not as stable as the vacuum type because the current will vary with variations in the applied voltage as well as with variations in light intensity. Stable voltage supplies must be used.

Another type of photocell is the photomultiplier, which is one of the most fascinating of modern scientific instruments. More sensitive than the human eye, it will often respond to individual photons. A schematic of its construction is shown in Figure 10-22 and its operation is as follows: Light ejects photoelectrons from the photocathode (C), and they are attracted to the first dynode (D_1),

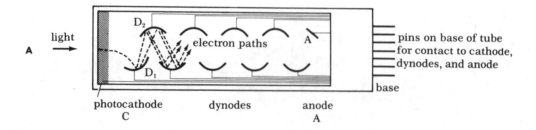

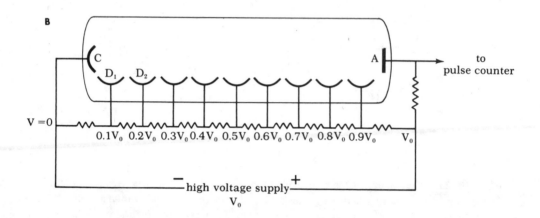

Figure 10-22 The photomultiplier tube. **A** is a pictorial drawing showing the semi-transparent photocathode, C; the dynodes – of which the first two, D_1 and D_2, are labeled; and the anode, A, which collects the electrons. Some of the initial electron paths are also shown. Diagram **B** is a schematic showing how each dynode is made more positive than the previous one by the use of a chain of resistors.

which is a hundred volts or so positive with respect to the photocathode (C). This dynode is coated with what is called an electron sensitive surface. When a photo-electron strikes it, five or more electrons are "splashed out," and these are attracted to the next dynode (D_2), where the process is repeated, giving 25 or more electrons. After nine such stages, if the average were five electrons ejected per stage, there would be 5^9 or about 2,000,000 electrons collected at the anode (A) for each photoelectron produced initially. In practice, amplifications of about a million are common and some photomultipliers will give even a hundred million electrons for each intial one. The photomultiplier, although it is very sensitive to light, is also very sensitive to voltage variations. The power supply, producing a total of a thousand volts or more, must be very steady.

A different class of photocell is that which is a completely solid state. Most of the satellites and space probes are powered by *solar cells,* which are devices to produce electricity from sunlight. They are really a solid state type of photocell. The history of this type of cell goes back a long way, but it has been only with the recent development of the silicon type, which uses the techniques evolved for the making of transistors, that they have reached their present status. The silicon cell consists of a very thin layer of silicon. Actually, this thin layer is two layers, for the top ten thousandth of an inch or so is p type silicon and below that is n

type silicon. Electrical contact is made to the n type by plating the back, and to the p type by plating a small strip on the front. Light falling on the cell gives energy to some of the bound electrons, again by the photoelectric effect, and a voltage appears across the cell. If the two sides of it are connected, a current results. The physics of this, I leave for the time of your later study of transistors, but you can use these cells for your experiments anyway. Unfortunately, except for very low illumination, the relation between current output and intensity of illumination is not linear but is logarithmic. A calibration curve is therefore absolutely necessary when these cells are used to compare light intensities.

DETECTORS OF X-RAYS AND GAMMA RAYS

X-rays and gamma rays are electromagnetic radiations like light but of a much higher energy or shorter wavelength. They are absorbed in a medium by processes called the photoelectric effect, the Compton effect, or pair production. The photoelectric effect involves the transfer of the total energy of the photon to an electron. This is in effect an inelastic collision, but in an inelastic collision (as you saw in Experiment 6-2) you can't have conservation of both momentum and energy. So for the photoelectric effect to occur, the electron must be bound to some object (an atom, perhaps), and the object to which it is bound carries away some of the momentum while almost all the energy is transferred to the electron. Some of this energy is lost in getting out of the atom, so we do not see it all. The Compton effect is effectively an elastic collision between a photon and a comparatively free electron. The photon transfers only part of its energy to the electron and continues on itself with reduced energy. This is similar to the collision studied in Experiment 6-3. In pair production, the photon, in the presence of a high electric field such as that near a nucleus, is transformed into a pair of particles, an ordinary negatively charged electron and a positively charged electron called a positron. The photon disappears. Positrons are an example of what is called antimatter. For radiation of energy over a thousand million electron volts (a Bev or Gev), a proton and antiproton pair may be produced. Such energies are produced only by our largest accelerators, and ordinarily the absorption process involves the production of only high speed electrons or electron-positron pairs. The detectors are devices to detect high speed particles. These devices can detect beta particles from radioactive materials also, since beta particles are only high speed electrons. They may be the negative or positive variety.

In your experiments you may use a Geiger counter or a scintillation counter, so both of these will be described. There are also many other devices to measure radiation, or really, to detect high speed charged particles.

The common Geiger counter, more correctly called a Geiger-Müller or G-M counter, consists of a fine wire down the center of a cylinder with a low pressure gas in between. Often a thin end window is built in to allow beta particles to enter and be counted.

A high voltage is put across the tube, with the center wire positive. The situation is somewhat analogous to the gas-filled photocell, but there is about a thousand volts across a G-M tube. The high speed electron itself, in passing through the gas, knocks electrons from the atoms of the gas and these secondary electrons are all attracted to the anode wire, producing more electrons as they collide with the atoms of the gas. So one high speed particle entering the tube produces a veritable avalanche of electrons at the center wire. The result is apparent as a brief pulse of current which can operate a counter or give the click associated with "Geiger" counters.

The experiments which are described using G-M counters could be done as well with scintillation counters. Some materials give off light when a high speed particle goes through them. The inside of the front of a TV screen is coated with just such a material that gives off a pulse of light for every electron that hits it. Radium watch dials give off a pulse of light from each alpha particle emitted by the radioactive material in the paint on the dial. You will be able to see the individual light pulses if you let your eyes get dark adapted (for at least five minutes) and then observe the watch dial carefully. A lens may help. The scintillation counter will frequently have a large crystal of such scintillating material. The gamma rays are absorbed in the crystal, producing a high speed particle and in turn a flash of light. The light flash is detected by a photomultiplier (see p. 209) and results in an electrical pulse which, as with a G-M tube, may be made to operate a counter. The more frequent the counts, the more intense the radiation.

COUNTERS

G-M or scintillation counters require a high speed counting device along with a timer in order to measure the radiation and express it in terms of a counting rate. The instruments to do this are of great variety and ingenuity but will almost invariably display the number of counts in an array of numbered lights. A switch will set the count to zero in preparation for another measurement. Another switch will start and stop the counting. The timing may be done automatically or with a clock. At any rate, when you are confronted with a counter, look for these switches and familiarize yourself with their operation.

10-1

The Compound Microscope

The object of this project is to set up a compound microscope consisting of two lenses on a vertical stand and to determine the magnifying power of the eyepiece, the objective lens, and the whole microscope.

A compound microscope consists of two sets of lenses (see Figure 10-7), an objective lens and an ocular or eyepiece. The objective lens forms a real, magnified image inside the tube of the microscope and usually 16 cm. from the objective lens. The eyepiece is used to examine this image exactly as one would examine a small object with a single lens. The total magnifying power is the product of the magnifications by the objective and eyepiece.

The magnifying power of a lens or simple microscope is defined as the ratio of the angular size of the object when viewed with the lens or microscope to the angular size when viewed at 25 cm. (or 10 inches) from the naked eye. In terms of the focal length, f, of a single lens, the magnifying power, $m.p.$, is given by

$$m.p. = \frac{25\,cm.}{f} + 1$$

When a lens is used to form a real image, the magnification is given by image size/object size because a real image is involved. It is also image distance/object distance.

In this project the eyepiece and the objective lens each will consist of a single lens. In actual microscopes each will be lens combinations to reduce various aberrations.

The project is in three parts: In the first part the magnifying power of a lens to be used as an eyepiece is determined. In the second part the magnification of a lens used as an objective lens is measured, and in the third part the magnifying power of the whole compound microscope is measured.

THE MAGNIFYING POWER OF THE EYEPIECE

A 10 cm. focal length lens is to be used for the eyepiece. Place the lens in a holder 25 cm. above a piece of paper with red concentric circles of radii 0.5, 1.0, 1.5, 2.0, and 2.5 cm., as shown in Figure 10-23A. Beneath the lens place another holder with the card having a black circle with a radius of 0.5 cm. The black circle is the object. Look through the lens and adjust the object up and down until a

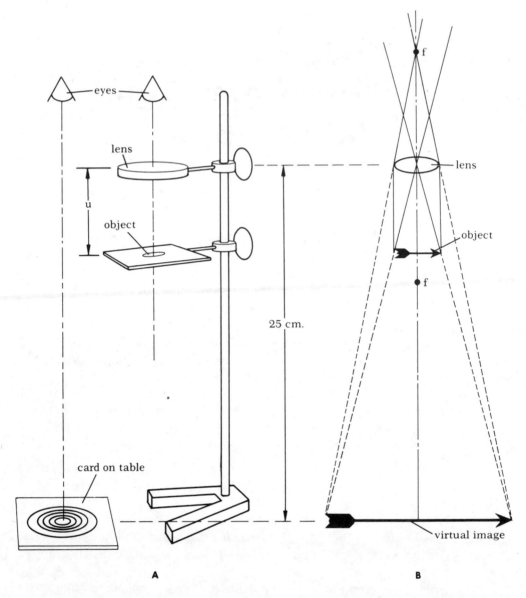

Figure 10-23 The use of a lens as a magnifier or simple microscope. A shows the physical arrangement and B shows the principal ray diagram. This arrangement is similar to the eyepiece of a compound microscope.

clear image is seen. Get this final image 25 cm. from the lens by using both eyes, one looking through the lens and the other at the paper on the table. The image of the black circle will be seen superimposed on the paper on the table. This is a visual "double exposure." The image will be at the same distance as the paper when there is no parallax between the black and the red circles. That is, as the eye is moved back and forth, the images move together.

Determine the magnifying power by comparing the sizes of the circles.

To see how well the system works, measure the object distance, u. The image distance, v, is -25 cm., the minus sign indicating that the image is on the same side of the lens as the object. Then calculate the focal length from $1/u + 1/v = 1/f$

and use the formula given above to calculate the expected *m.p.* Compare the two results.

THE MAGNIFYING POWER OF THE OBJECTIVE LENS

A lens with a focal length about 5 cm. makes a satisfactory objective lens for this project. The object can be a small hole behind which is a light bulb. The object size, which is the diameter of the hole, can be measured with a vernier caliper. The image size can be measured with a transparent glass scale. To measure the *m.p.* of the objective lens, arrange the apparatus on the stand, as shown in Figure 10–24. The glass scale should be put at the same position on the stand as the object was for the first part of the experiment, and the objective lens should be 16 cm. below this.

The object then is moved up and down until a sharp image is formed on the glass scale. To obtain the best focus and also to help read the image size on the scale, the eyepiece lens is placed at a distance u above the glass scale, u being the same as in the first part of this experiment. Note and record the size of the image of the hole on the glass scale.

What magnification was produced by the objective lens?

Measure the object and image distances and calculate from them the expected magnification. Does the answer compare satisfactorily with that found by direct measurement?

THE MAGNIFYING POWER OF THE MICROSCOPE

The arrangement finally arrived at in the second part of the experiment is a complete compound microscope as shown in Figure 10–25. The objective lens produces a magnified image at the position of the scale, and the eyepiece is used to view this image and give further magnification. The total magnification is the product of those from the objective and the eyepiece. Calculate this from the results of the first two parts of the experiment.

Try the microscope by turning out the light under the object and placing an object such as a stamp at the object position. Measure the magnifying power by using a ruler as an object and employing the same technique you used in finding the magnifying power of the eyepiece.

If time is available, calibrate the divisions on the glass scale by using a ruler for an object. Then use your microscope to measure a small object. Record all the data.

Apparatus

Lab stand about 50 cm. high
4 right angle clamps
2 lens holders
1 holder for glass scale (perhaps another lens holder)
Millimeter scale etched on glass
Lens of focal length 10 cm.

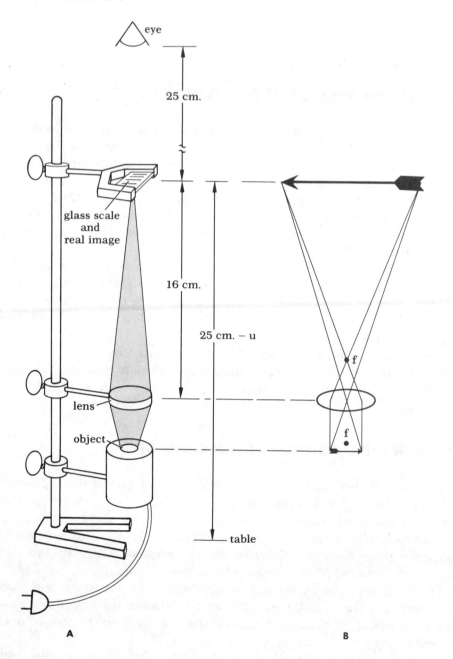

Figure 10-24 The use of a lens to make a real image. **A** shows the physical arrangement and **B** shows the principal ray diagram. This arrangement is similar to the objective lens of a compound microscope.

Lens of focal length 5 cm.
Small lamp (7 1/2 watt) in housing with 1-2 mm. diameter hole
Small card with 0.5 cm. radius black circle
Card with red circles of 0.5, 1.0, 1.5, 2.0, and 2.5 cm.
30 cm. scale or meter scale
Vernier caliper
Small object such as a stamp

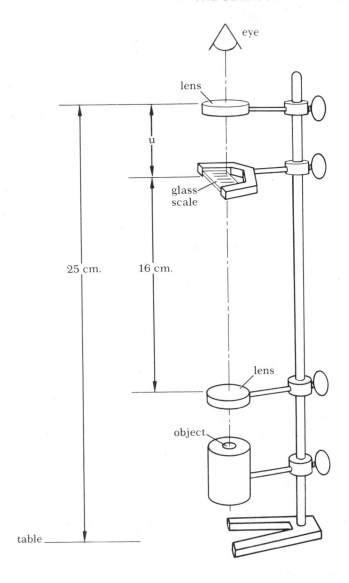

Figure 10-25 The compound microscope. This is a combination of the arrangements shown in Figures 10-23 and 10-24. The image on the glass scale acts as the object for the eyepiece.

10-2

The Stroboscope

You have used the stroboscope in several of your experiments, but how does it work? It gives regular pulses of light of very short duration from the gas discharge lamp, but how does it accomplish this? What is the way in which the frequency is made to change? The purpose of this experiment is to demonstrate how the strobe lamp works and to show some of the rather unusual electrical properties of gas-filled tubes.

First you are to do an Ohm's law experiment with a neon bulb; that is, plot a curve of current against voltage. To do this use the circuit shown in Figure 10-26. The voltage source is continuously variable up to 150 volts. The voltmeter is connected so that the current to the meter is not registered by the milliammeter. The resistance, R, is for protection; that is, it limits the current. With this arrangement, the voltmeter reads the voltage V_o provided by the power supply but does not read the voltage (V) across the neon bulb. By Ohm's law, if a current i flows through a resistance R, there is a voltage drop equal to iR in the resistance. So the voltage V is given by $V_o - iR$. To take your readings gradually, raise V_o to its maximum and then gradually lower it without going back to repeat readings. Tabulate your data in a table with column headings of V_o, i, iR, and V. Then graph i against V_o. Note what is called the firing voltage.

You can raise the voltage slowly and automatically by using the circuit shown in Figure 10-27. The bulb is across the condenser C, and the condenser charges slowly through the resistance, R. When the firing voltage is reached, the bulb flashes, the condenser loses some of its charge so that the voltage goes down, and the discharge stops. The condenser begins to charge again and the process repeats, giving a periodic flash.

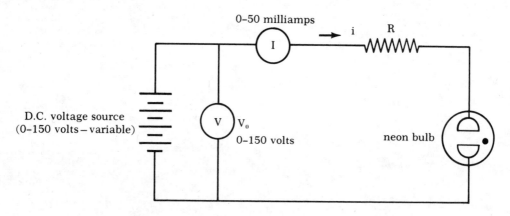

Figure 10-26 The circuit to study current-voltage relations for a neon bulb.

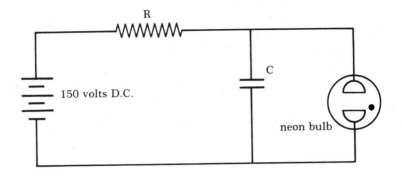

Figure 10-27 The circuit for a simple strobe lamp.

Investigate how the time between flashes changes with different values of R and C.

Make a written report on this experiment.

Apparatus

Neon bulb — NE2
Voltmeter — 0-150 volts (or multimeter)
Resistances — 1, 5, 10 megohms
D.C. current meter reading to 50 mA (or multimeter)
Capacitors — 0.5, 1, 2 μF, 400 volts
150 volt D.C. supply continuously variable

10-3

Measurement with Waves of Light

The purpose of this project is to determine the number of waves of light that correspond to the diameter of a human hair or to the thickness of a metal foil.

The method to be used is that described on page 193 using two optically flat glass plates about 4 inches long and 1 inch wide. The two glass plates are set on a black cloth and are separated at one end by a hair obtained from one of the students. The hair will be easier to handle if it is glued across a V-shaped piece of cardboard. A sodium lamp or flame, or a mercury lamp with a green filter, may be used for illumination. An oblique glass plate is used to reflect the light down onto the apparatus, and the fringes (as shown in Figure 10–11) should be observed. They can be most easily observed in a slightly darkened room. The problem is to determine the number of fringes along the plates.

The obvious method is to count the fringes and this can be done after some practice — but not without confusion and recounting. If this method is used, several counts should be made, if possible, by different observers. Then, by reference to page 194 for the theory, you can find the number of waves in the thickness of the hair.

A more reliable way to count the fringes is to view them with a traveling microscope and then use the cross hairs of the microscope to keep your position while counting.

After the number of waves in the width of the hair has been determined, some interesting things may be done. The first is to measure the hair with a micrometer caliper and from this measurement calculate the wavelength of the light. The second is to look up the wavelength of the light used and calculate the diameter of the hair. Do one or the other or both and make comments about the precision.

A piece of carefully flattened aluminum foil or shim stock may be substituted for the hair in this project. The experiment could also be repeated using hair from another student, perhaps comparing the diameters of the hairs from blonds, brunettes, etc.

Water may be put on a small area between the plates by touching a drop on a glass rod to the edge of the air space. You can then tell by the fringe spacing whether the waves are shorter or longer than those in air. This indicates also, from $v = f\lambda$, whether light travels faster or slower in water.

Apparatus

2 optically flat glass plates 1 inch × 4 inches
1 hair
Monochromatic light source (sodium lamp or mercury lamp with a filter)
Measuring microscope (optional)
Micrometer caliper (optional)
Stands and holders
Black card or cloth

10-4

Meter Resistance

VOLTMETER

Many voltmeters do not have marked on them either their resistance or their sensitivity in ohms per volt, so it may be necessary to measure the meter resistance. The purpose of this project is to determine the internal resistance of a voltmeter. You will use a voltage source which gives approximately full scale deflection of the meter and a series of known resistances covering wide range so that one which is roughly the same as the meter resistance can be found.

The circuit diagram is shown in Figure 10-28. R is a resistance variable by substitution. When R is zero, the meter reading V_1 is equal to the voltage of the source. Then various values of R are tried until one is found which cuts the meter reading to value of V_2 which is approximately $V_1/2$. The circuit resistance is the meter resistance plus R, the internal resistance of the battery being negligible.

To solve for the meter resistance R_m, Ohm's law is applied in the two cases: for $R = 0$ and then for the value of R finally found to give an appreciable voltage drop. In this way it can be shown, and is left for you to show, that the voltmeter resistance, R_m, is given by

$$R_m = RV_2/(V_1 - V_2)$$

To show this, keep in mind that the voltmeter actually indicates a deflection proportional to the current passing through it. The ratio of currents i_1 and i_2 for situations which give indicated voltage readings V_1 and V_2 is the same as V_1/V_2.

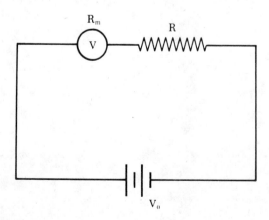

Figure 10-28 The circuit used to find the resistance of a voltmeter.

You can write expressions (using Ohm's law) for i_1 and i_2, the currents in your two circuits, one with no added resistance ($R = 0$) and the other with R not zero. Then find the ratio i_1/i_2 which will be equal to the ratio of your indicated voltage readings. What is the sensitivity of the meter in ohms per volt?

CURRENT METER

Current meters have relatively low resistance, so in general it is not possible to put a voltage source such as a battery directly across them: The current would be too much for the meter. A series resistance calculated to limit the current within the range of the meter must be added.

To determine the meter resistance, use a voltage source and series resistance calculated to give approximately full scale meter deflection. (The circuit is shown in Figure 10–29A.) Then put a shunt resistance across the meter as shown in Figure 10–29B and vary this until the current reading is reduced to about one half. Using the values of the resistances in both cases and the two current readings, calculate the meter resistance. If I_1 is the indicated current when there is no shunt resistance and I_2 is the indicated current with a shunt resistance R, then you are to show that the meter resistance, R_m is given by

$$R_m = R(I_1 - I_2)/I_2$$

Apparatus

Voltmeter
Ammeter, milliammeter, or microammeter
Resistance boxes — one of high range and one of low range (chosen to be appropriate for the meters)
Voltage sources and controlling rheostats

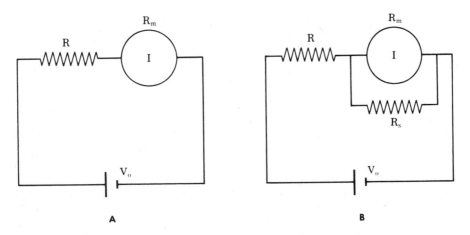

Figure 10–29 The circuits used to find the resistance of a current meter (ammeter).

REFERENCES

1. Lawrence, W.L.: The miracle that saved the world. *In* Rapport, S., and Wright, H., eds: Science: Method and Meaning. New York, New York University Press, 1963, p. 179.

Name _____

Partner _____

Date _____

Summary of data:

Degrees θ	Number, n, of oscillations	Time for n oscillations	Time for one oscillation

Conclusions:

Name _____

Partner _____

Date _____

A. The mean ratio, real depth to apparent depth, for water was _____.

B. The ratio, real thickness to apparent thickness, for glass in air was

_____.

C. The ratio, real thickness to apparent thickness, for a glass block under a

layer of water was _____ .

D. The theoretical relation between real depth and apparent depth was

_____ .

Comments:

REPORT SHEET

EXPERIMENT 2-2

OHM'S LAW

Name _____

Partner _____

Date _____

A. Carbon or wire-wound resistor. (A graph is attached.)

The rated value of R was _____ ohms.

V, Volts	I, Predicted	I, Measured	$R = V/I$, Measured
0			
2			
4			
6			
8			
10			

The average value of R from column 4 is _____ ohms.

B. The data for the other objects that were investigated are shown on the accompanying graph. Comments on each are below.

REPORT SHEET
EXPERIMENT 2-3
THE BOUNCING BALL

Name _____

Partner _____

Date _____

Summary of Data:

Trial A   Trial B

Material of ball _____   Material of ball _____

Material of surface _____   Material of surface _____

h_1	h_2	v_1	v_2	e

h_1	h_2	v_1	v_2	e

Comments:

Number of bounces the ball will make is _____ .

Series Equation for time is _____ .

Total time for ball in A to come to rest after falling from height

h_1 is _____ (theoretical)

_____ (measured).

REPORT SHEET

EXPERIMENT 2–4

LENSES AND IMAGES

Name _____

Partner _____

Date _____

A set of measurements, and in the third column, the combination of a and b that yields a constant:

a	b	

The equation which is the combination of a and b that yields a constant is

_____ .

The approximate measured value of f is _____ .

The probable equation relating a, b and f is _____ .

Name _____

Partner _____

Date _____

Distance s = _____ meters ± _____ %.

Height h = _____ meters ± _____ %.

Calculated time T = _____ sec. ± _____ sec.

Measured times:

	Sphere		Solid cylinder		Hollow cylinder

t	$t - \bar{t}$	t	$t - \bar{t}$	t	$t - \bar{t}$

Summary of results:

Calculated T = _____ sec. ± _____ .

For sphere $\bar{t}$ = _____ ± _____ .

Ratio to calculated time _____ .

For solid cylinder $\bar{t}$ = _____ ± _____ .

Ratio to calculated time _____ .

For hollow cylinder $\bar{t} =$ _____ $\pm$ _____ .

Ratio to calculated time _____ .

You should submit a conclusion in a separate paragraph.

Name _____

Partner _____

Date _____

Indicated value of the resistors = _____ ohms.

Indicated tolerance = _____ %.

The comparison resistance in the bridge, R_2, was _____ ohms.

The quantities l_1, l_2, and R_2 are described in Chapter Ten.

Resistor Number	l_1	l_2	Measured Value, R	$R - \bar{R}$	$(R - \bar{R})^2$
Sums					

Mean value = _____ ohms.

Standard deviation = _____ ohms = ± _____ %.

The difference between the mean and the indicated value was _____ ohms

or _____ %.

What fraction of the resistors was outside the specified range? _____

Name _____

Partner _____

Date _____

The results are summarized below:

n	$\bar{x}$	S^2	S

Conclusion:

Between $\bar{x} - S$ and $\bar{x} + S$ there was _____ % of the readings.

There was _____ % of the readings outside the range $\bar{x} \pm 2S$.

Answers to questions:

1.

2.

3.

REPORT SHEET
EXPERIMENT 5–1
NEWTON'S RINGS

Name _____

Partner _____

Date _____

Ring Number *n*	Microscope Readings		Diameter in mm.	Radius in mm., *r*	Log *r*	Log *n*
	Left	*Right*				

Results:

The explanation of the rings:

REPORT SHEET
EXPERIMENT 5-2
FALLING OBJECTS

Name _____

Partner _____

Date _____

Flashing rate of strobe lamp = _____ .

Time interval between flashes = _____ sec.

Conversion factor from film to space: 1 mm. on film = _____ in space.

The results of the measurement of the photograph are below and a graph of one of these sets of data is attached.

t in sec.	s in meters	v in meters/sec.

Answers to the questions:

1.

2.

3.

4.

Name _____

Partner _____

Date _____

Scale factor: 1 mm. on film = _____ in space.

Flashing rate of stroboscope = _____ .

Interval between flashes   = _____ sec.

Tables of data and calculated results:

Dot Number	Microscope Readings		Distance from First Dot				Time of Dot
	Horizontal	Vertical	On Film		In Space		
			x	y	x	y	
1							0

Analysis:

Interval Number	Δx	Δy	v_x	v_y	t
1					0

Results:

These will include two graphs, one showing the horizontal distance x and the vertical distance y, each as a function of time, and the other showing the velocities v_z and v_y, each as a function of time. From the graphs the following relations were obtained:

$x = f(t) = $ _____ $v_y = f(t) = $ _____

$v_x = f(t) = $ _____ $a_y = $ _____

$a_x = $ _____

Conclusions (include generalizations from the above relations):

REPORT SHEET

EXPERIMENT 5-4

MOMENT OF INERTIA OF A SOLID DISC

Name _____

Partner _____

Date _____

Radius of hub or cylinder (including half the string), $r =$ _____ .

Disc Number	Radius of Disc, R	Mass of Disc, M
1		
2		
3		
4		
5		

Average mass = _____ .

Summary of measurements:

Disc. Number	Distance Fallen, s	Time to Fall, t	Acceleration, a	$g - a$
none				
1				
2				
3				
4				
5				

Disc Number	Falling Mass, m	$T = m(g - a)$	Torque, $L = Tr$	$\alpha = a/r$	L/α	$I = L/\alpha - I_o$
none						
1						
2						
3						
4						
5						

Results of analysis: $I = f(r) =$ _____ .

The general formula for the moment of inertia of a solid disc is probably:

$I =$ _____ .

REPORT SHEET

EXPERIMENT 5-5

TENSION IN A CORD ON A CYLINDER

Name _____

Partner _____

Date _____

Data:

θ	T	T_o	T_o/T

Analysis of the data showed that T was related to T_o and θ by:

$$T = \underline{\hspace{3cm}}.$$

Optional: Theoretical analysis yielded:

$$T = \underline{\hspace{3cm}}.$$

The coefficient of friction between the cord and the cylinder was _____ .

REPORT SHEET
EXPERIMENT 5-6
FLUID FLOW IN TUBES

Name _____

Partner _____

Date _____

Radius, r	Quantity, Q	Time, t	Q/t

A graph is attached.

The relation between rate of flow and radius was found from the graph to be:

$$Q/t = \underline{\hspace{2cm}} .$$

The basic form is probably:

$$Q/t = \underline{\hspace{2cm}} .$$

REPORT SHEET

EXPERIMENT 5-7

LIGHT FROM A LINEAR SOURCE

Name _____

Partner _____

Date _____

Perpendicular Distance, d	Photocell Current, μA	Relative Intensity, E

A straight line graph of the data is attached.

The relation between relative intensity E and distance d was found to be:

$$E = \text{\underline{\hspace{3cm}}} \, .$$

The form of the relation is probably:

$$E = \text{\underline{\hspace{3cm}}} \, .$$

EXPERIMENT 5-8

FIELD NEAR A MAGNETIC DIPOLE

Name _____

Partner _____

Date _____

Data and analysis:

The angles θ_1 and θ_2 are the compass deflections for the compass being located on one side and then on the other side of the bar magnet. The mean deflection is $\bar{\theta}$.

x	θ_1	θ_2	$\bar{\theta}$	$\tan \bar{\theta} = B_m/B_e$

A graph (of a type which yields a straight line) is attached.

Conclusions:

The equation relating the field strength, as represented by B_m/B_e, to the distance from the dipole was found to be:

$$B_m/B_e = \text{\underline{\hspace{3cm}}} \; .$$

The general form of the relation is probably:

$$B_m/B_e = \text{\underline{\hspace{3cm}}} \; .$$

REPORT SHEET

EXPERIMENT 5-9

A CAPACITOR IN AN A.C. CIRCUIT

Name _____

Partner _____

Date _____

The voltage source is _____ volts and _____ cycles per sec.

Table of data for the effect of change of capacity:

Capacity, C	Voltage, V (M_1)	Current, I (M_2)	Reactance, X_c Ohms

A graph is attached and the relation expressing X_c in terms of C was actually:

$$X_c = \text{\underline{\hspace{3cm}}} \, .$$

The general form is probably:

$$X_c = \text{\underline{\hspace{3cm}}} \, .$$

REPORT SHEET

EXPERIMENT 5-10

RESONANCE, CAPACITY, AND INDUCTANCE

Name _____

Partner _____

Date _____

Data:

Inductance constant at $L =$ _____ .   Capacity constant at $C =$ _____ .

Capacity, C	Resonant Frequency, f

Inductance, L	Resonant Frequency, f

A graph is attached showing the relation of resonant frequency to capacitance, and the relation is actually:

$$f = \text{_____} .$$

A graph is attached showing the relation of resonant frequency to inductance, and the relation is actually:

$$f = \text{_____} .$$

The general relation seems to be:

$$f = \text{_____} .$$

A discussion (including theoretical analysis for the constant) should be submitted on a separate sheet.

259

REPORT SHEET

EXPERIMENT 5-11

THE LAW OF GROWTH AND DECAY

Name _____

Partner _____

Date _____

In each part of this experiment you were to make an estimate of the value of K in the relation which is in the form $\Delta x/\Delta t = Kx$. In each case the value of Δx will have a large error but an estimate of K will have been made. Then the relation between x and t was to have been found for each case and the position of the K determined to give you the general form of the relation. So for each experiment you did, report the equations you found, the value of K, and the general relation suspected from all of your experiments.

1. Title of experiment: _____ .

    The equation $\Delta x/\Delta t = Kx$ took the form _____

    for this experiment with $K =$ _____ .

    The relation between the quantities represented by x and t was:

    $x =$ _____ .

2. Title of experiment: _____ .

    The equation $\Delta x/\Delta t = Kx$ took the form _____ with

    $K =$ _____ .

    The relation between the quantities represented by x and t was:

    $x =$ _____ .

3. Generalization:

    For situations described by $\Delta x/\Delta t = Kx$, the relation between x, t, and K was

    found to be given by: $x =$ _____ .

REPORT SHEET

EXPERIMENT 6–1

RESOLVING POWER

Name _____

Partner _____

Date _____

A.   A description of the phenomenon:

B.   Data showing combinations of R.P. and of *a* that are not constant and a
      combination that is constant.

R.P. (Radians)	*a* (mm.)			

      Mean value of the constant, *k*, is _____ .

C.   With R.P. in radians, *a* in mm. and λ in mm., the relation between the
      constant and λ is *k* = _____ .

D.   The complete R.P. equation in terms of radians, wavelength and aperture is

      R.P. = _____ .

E.   The expected R.P. of the 200-inch telescope, using the value of *k* found

      here, is _____ radians = _____ sec.

      The expected resolution of the 250-foot and 3300-foot radio telescopes for
      21-cm. waves, on the basis of the results of this experiment would be

      _____ and _____ .

F.   The resolving power of my eye is _____ minutes.

REPORT SHEET

EXPERIMENT 6-2

INELASTIC COLLISIONS AND MUZZLE VELOCITY

Name _____

Partner _____

Date _____

Mass:

Mass of moving car m_1 = _____ kg.

Mass of stationary car m_2 = _____ kg.

Velocity:

Velocity v_1 of m_1 just before collision = _____ meters/sec.

Velocity v_2 of $(m_1 + m_2)$ just after collision = _____ meters/sec.

Momentum:

Momentum before the collision $m_1 v_1$ = _____ kg. meters/sec.

Momentum after the collision $(m_1 + m_2)v_2$ = _____ kg. meters/sec.

Kinetic energy:

K.E. before the collision $1/2\, m_1 v_1^2$ = _____ joules.

K.E. after the collision $1/2(m_1 + m_2)v_2^2$ = _____ joules.

A graph of velocity against time is attached.

Conclusion:

Muzzle velocity:

   Mass of projectile m_1 = _____ kg.

   Mass of block and cart m_2 = _____ kg.

   Velocity after firing v_2 = _____ meters/sec.

   Velocity of projectile v_1 = _____ meters/sec.

Comments:

Name _____

Partner _____

Date _____

Flashing rate of strobe lamp   =

Mass of projectile m_1 =

Mass of target m_2 =

Velocity of projectile before collision v_1 =

Velocity of target before collision v_2 =

Velocity of projectile after collision v_3 =

Velocity of target after collision v_4 =

Momenta: $m_1 v_1$ =

                        $m_1 v_3$ =

                        $m_2 v_4$ =

Vector sum of momenta after collision   =

Kinetic energy: $m_1 v^2{}_1/2$ =

                $m_1 v^2{}_3/2$ =

                $m_2 v^2{}_4/2$ =

Total kinetic energy after collision   =

Velocity of separation   =

Velocity of approach   =

Coefficient of restitution $e =$

Compare the total momenta before and after the collision and the total K.E. before and after. Comment on your findings.

Name _____

Partner _____

Date _____

Data and calculations:

Frequency f	Number of Half Waves Measured	Measured Distance, $n\lambda/2$	Wavelength, λ	$v = f\lambda$

Summary:

f	v

Conclusions, with error:

REPORT SHEET
EXPERIMENT 7-1
SIMPLE HARMONIC MOTION

Name _____

Partner _____

Date _____

Spiral Spring

F	x

Force constant k = _____ $\pm$ _____ .

Mass m to be vibrated = _____ $\pm$ _____ .

Expected period = _____ $\pm$ _____ .

Measured period = _____ $\pm$ _____ .

Comment:

Metal Bar

F	x

Force constant k = _____ $\pm$ _____ .

Mass m to be vibrated = _____ = _____ .

Expected period = _____ $\pm$ _____ .

Measured period = _____ $\pm$ _____ .

Comment:

Summary of the results when the effective mass of the bar is considered:

REPORT SHEET
EXPERIMENT 7-2
THE DOPPLER EFFECT

Name _____

Partner _____

Date _____

Source frequency, f_s = _____ cycles/sec.

Number of beats per sec., $f_o - f_s$ = _____ /sec.

Temperature, T = _____ °C.

Velocity of sound, V = _____ meters/sec.

Solution for the speed, v:

The velocity of source, v, calculated on the basis of the Doppler effect was

= _____ meters/sec.

Distance over which the moving speaker was timed, s = _____ meters.

Time to travel the distance s = _____ sec.

Velocity, v = _____ meters/sec.

Comparison of results of the two methods:

REPORT SHEET

EXPERIMENT 7-3

THE SPEED OF ELECTRONS

Name _____

Partner _____

Date _____

A. A graph of number of electrons (expressed in microamps) in each speed interval is attached.

B. The data follow:

Speed v at Center of Interval	Number of Electrons N, Microamps	v^2	$N v^2$
$\times 10^4$ m./sec.			

C. The r.m.s. speed of the electrons, $\sqrt{\Sigma\, Nv^2 / \Sigma\, N} =$ _____ m/sec.

D. The temperature associated with this speed is _____ °K.

E. The r.m.s. speed of air molecules at that temperature is _____ m/sec.

$=$ _____ mi./hr.

EXPERIMENT 8-1

AIR THERMOMETER

Name _____

Partner _____

Date _____

Barometric pressure, $B =$ _____ cm. of Hg.

Data:

	h	$P = h + B$	$T°K$
Ice water			273.16
Room temperature			
Dry ice (solid CO_2)			

The temperature of the dry ice was _____ °C.

With the blackened bulb a foot away from a hundred-watt lamp, the temperature

rose to _____ °K.

REPORT SHEET

EXPERIMENT 8–2

THE DETERMINATION OF g BY MEANS OF A SIMPLE PENDULUM

Name _____

Partner _____

Date _____

Length, l = _____ ± _____ .

Measurement of period:

Number of Swings	Time	Period
Sums:		

Mean period, T = _____ ± _____ .

Measured value of $g = 4\pi^2 \ l/T^2$ = _____ ± _____ .

Theoretical value of g = _____ .

Discussion of result:

EXPERIMENT 8–3

THE PHYSICAL PENDULUM TO MEASURE g

Name _____

Partner _____

Date _____

Type of pendulum used: _____.

Identification of watch: _____.

Calibration data for watch: _____.

The following lengths were measured: _____.

Numer of Swings	Time	Period, T

Mean period, T = _____ sec. ± _____ sec.

Corrected for calibration (if applicable), T = _____ sec. ± _____ sec.

Result: Measured value of g = _____ ± _____ .

Theoretical value of g = _____ .

Comments:

REPORT SHEET
EXPERIMENT 8–4
TORSION PENDULUM

Name _____

Partner _____

Date _____

1. Calculation of k:

   l = _____ cm. = _____ meters.

   r = _____ = _____ meters.

   $E_s = 8 \times 10^{10}$ nt./m.^{2}

   k = _____ .

2. Direct measurement of k:

   Mass on holder = _____ kg. ± _____ .

   mg = _____ newtons ± _____ .

   Torque L = _____ newton meters ± _____ .

   $\theta = 2\pi$ radians

   $k = L/\theta$ _____ .

3. Torsion pendulum:

   Number of oscillations = _____ .

   Time = _____ ± _____ .

   Period = _____ ± _____ .

   M = _____ kg. ± _____ .

   R = _____ meters ± _____ .

   I = _____ .

   k = _____ .

Summary:

by calculation $k =$ _____ .

by direct measurement $k =$ _____   ±   _____ .

by torsion pendulum $k =$ _____   ±   _____ .

Comment:

Name _____

Partner _____

Date _____

Material of the wire: _____

Length l = _____ meters

Measurements of diameter: Measurement of elongation:

Number	d

Added Mass, m	Microscope Readings			e
	1	2	3	
0				0

Mean diameter = _____ meters ± _____ .

Calculated values of Y:

Mean value of Y = _____ ± _____ .

Value of Y from Tables:

Discussion of results:

Name _____

Partner _____

Date _____

Prism angle: _____°.

Measurement of minimum deviation:

Color	Angle of Undeviated Light	Angle of Deviated Light	Angle of Deviation, D

Summary of index of refraction calculations:

Color	Index, μ

Discussion of the error in μ and of the type of glass:

REPORT SHEET
EXPERIMENT 8-7
THE UNIVERSAL GRAVITATION CONSTANT, G

Name _____

Partner _____

Date _____

Data for the calculation of the force between the balls:

Distance of mirror to scale, D = _____ meters ± _____.

Equilibrium position for configuration **A** = _____ ± _____.

Equilibrium position for configuration **B** = _____ ± _____.

Change in equilibrium position, s = _____ meters ± _____.

Angular change $s/D = 4\theta$ = _____ radians ± _____.

Angular displacement due to gravitational force (between only one pair of balls),

  θ = _____ radians ± _____.

Radius of small ball, a = _____ meters ± _____.

Distance of small ball to fiber, b = _____ meters ± _____.

Moment of inertia, I = _____ kg. meters^2 ± _____.

Period of oscillation, T = _____ seconds ± _____.

Torsion constant, k = _____ nt.m/radian ± _____.

Torque which caused the displacement θ, = _____ newton meters ± _____.

This is the gravitational force between one pair of balls of masses m_1 and m_2 separated by the distance r.

The universal gravitational constant, G, is then found from:

$$F = \frac{Gm_1 m_2}{r^2}$$

F, the gravitational force = _____ newtons ± _____ .

m_1, the mass of a large ball = _____ kilograms ± _____ .

m_2, the mass of a small ball = _____ kilograms ± _____ .

r, the distance between their centers = _____ meters ± _____ .

Resulting in:

G = _____ ± _____ .

Name _____

Partner _____

Date _____

Color of Filter	Wavelength	Stopping Voltage	Maximum Energy of Electrons

Planck's constant was found to be: _____ .

The energy carried by individual photons as found for the different colors:

Color	Energy in One Photon

Name _____

Partner _____

Date _____

Description of object:

Mean weight in air   = _____ gm.

Mean weight in water = _____ gm.

Mean loss in weight   = _____ gm.

Mean s.g.   = _____ .

Suspected materials: (a) _____ s.g. = _____ .

                      (b) _____ s.g. = _____ .

Suspected composition of object:

REPORT SHEET
EXPERIMENT 9-2
SPECTROSCOPIC ANALYSIS

Name _____

Partner _____

Date _____

Number of lines per inch on grating = _____

Line spacing, s = _____ nm. or mμ

| Color | Brightness | Protractor Readings: | | 2θ | θ | $s \sin \theta = n\lambda$ | n | λ |
		Right	Left					

Suspected element _____

Measured Wavelengths	Corresponding Wavelengths from Table 9-1 and 9-4

Conclusions:

The limit of my vision at the red end of the spectrum was _____ ,

and at the violet end it was _____ .

REPORT SHEET
EXPERIMENT 9-3
SPECTRAL ANALYSIS WITH A CAMERA

Name _____

Partner _____

Date _____

Focal length of camera lens = _____ mm.

Distance to source   = _____ meters.

Lens-to-film distance   = _____ mm.

Lines per inch in grating   = _____ .

Line spacing, s,   = _____ nanometers.

    Each unknown object is described with the measured wavelengths, and beside each is the corresponding wavelength of the suspected element.

(a)   Object was: _____   (b)   Object was: _____

       Element is: _____   Element is: _____

Measured Wavelengths	Accepted Wavelengths

Measured Wavelengths	Accepted Wavelengths

REPORT SHEET
EXPERIMENT 10-1
COMPOUND MICROSCOPE

Name _____

Partner _____

Date _____

Eyepiece

Radius of black circle = _____ mm.

Radius of image = _____ mm.

Measured $m.p.$ = _____ X

Object distance u = _____ cm.

Image distance v = _____ cm.

Focal length f = _____ cm.

Calculated $m.p.$ = _____ X

Comment:

Objective Lens (image distance = 16 cm.)

Diameter of object = _____ mm.

Diameter of image = _____ mm.

Measured magnification = _____ X

Object distance = _____ cm.

Image distance = _____ cm.

Calculated magnification = ___ X

Comment:

Compound Microscope

Calculated magnifying power = _____ X

Magnifying power found by looking at a scale:

Object size = 1 mm. spacing on scale.

Image size = _____ mm.

Measured $m.p.$ = _____ X

Comment:

Name _____

Partner _____

Date _____

A description of the current – voltage relation taken from the data in the lab record book follows, with a sketch of the curve and the numerical value of the firing voltage.

The way in which the firing rate varied with different values of C was:

The way in which the firing rate varied with different values of R was:

Name _____

Partner _____

Date _____

Object being measured: _____

Type of light used: _____

Wavelength of light used: _____

Measurement of fringes: number _____

distance _____

Length of plate to separating object: _____

Number of waves in the thickness of the object: _____

Calculated thickness of the object: _____

Measured thickness of the object: _____

Comments:

REPORT SHEET
EXPERIMENT 10-4
METER RESISTANCE

Name _____

Partner _____

Date _____

Voltmeter

Description of meter: _____

Voltage V_1 = _____

Voltage V_2 = _____

Added resistance, R = _____

Meter resistance, R_m = _____

Full scale reading = _____

Sensitivity in ohms per volt = _____

Current Meter

Description of meter: _____

Current, I_1 = _____

Current, I_2 = _____

Shunt resistance, R = _____

Meter resistance, R_m = _____

1

APPENDIX

THE BINOMIAL THEOREM AS IT APPLIES TO $(1+x)^p$

By multiplication it can be shown that if p is an integer, $(1 + x)^p$ can be put into the form of a finite polynomial.

$$(1 + x)^p = 1 + C_1 x + C_2 x^2 + \ldots x^p,$$

and the constants C_1, C_2, etc., can easily be evaluated.

We will make the assumption that $(1 + x)^p$ can be represented by a polynomial, perhaps infinite, whether or not p is integral. This will be assumed to be valid if the constants can be evaluated.

Let $f(x) = (1 + x)^p$, and assume $f(x) = 1 + C_1 x + C_2 x^2 + C_3 x^3 + C_4 x^4 + C_5 x^5 + \ldots$.

Differentiating the polynomial, $f'(x) = C_1 + 2C_2 x + 3C_3 x^2 + 4C_4 x3 + 5C_5 x^4 + \ldots$.

Evaluating this at $x = 0$, $f'(0) = C_1$.

Differentiating $f(x) = (1 + x)^p$

$$f'(x) = p (1 + x)^{(p-1)}$$

so

$f'(0) = p$, but $f'(0)$ has already been shown to be C_1.

The constant C_1 is just p.

Differentiating $f'(x)$ as obtained from the polynomial

$$f''(x) = 2C_2 + 2 \cdot 3C_3 x + 3 \cdot 4C_4 x^2 + 4 \cdot 5C_5 x^3 + \ldots.$$

$$f''(0) = 2C_2$$

Differentiating $f'(x)$ as obtained from $f(x) = (1 + x)^p$

$$f''(x) = p(p-1)(1+x)^{(p-2)}$$

and at $x = 0$, $\quad f''(0) = p(p-1)$

therefore $\quad 2C_2 = p(p-1)$

or $\quad C_2 = \dfrac{p(p-1)}{2}$

Differentiating a third time

$$f'''(x) = 2 \cdot 3 C_3 + 2 \cdot 3 \cdot 4 C_4 x + 3 \cdot 4 \cdot 5 C_5 x^2 + \ldots$$

$$f'''(0) = 2 \cdot 3 C_3$$

also $\quad f'''(x) = p(p-1)(p-2)(1+x)^{(p-3)}$

and $\quad f'''(0) = p(p-1)(p-2)$

therefore $\quad 2 \cdot 3 C_3 = p(p-1)(p-2)$

or $\quad C_3 = \dfrac{p(p-1)(p-2)}{2 \cdot 3}$

The general expression for the constant C_n becomes

$$C_n = \frac{p!}{n!(p-n)!}$$

The expression for $(1+x)^p$ then becomes

$$(1-x)^p = 1 + px + \frac{p(p-1)}{2} x^2 + \cdots \frac{p!}{n!(p-n)!} x^n + \cdots$$

This expression can be very useful in physics, especially in the situations in which the quantity x is much less than one. Then the term in x^2 and all terms with higher powers of x will be small enough to be neglected. It is then possible to use

$$(1 + x)^p = 1 + px \text{ if } x \ll 1$$

An expression often encountered in a problem is $(a + b)^p$, with $b \ll a$. This can be written as

$$(a + b)^p = a^p (1 + b/a)^p$$

Letting b/a be x, this is the same form as was dealt with above, and it becomes

$$(a + b)^p = a^p (1 + pb/a) \text{ if } b/a \ll 1$$

For example, evaluate $(1.01)^{10}$. Evaluating this gives, to six figures, $(1.01)^{10}$ = 1.10462. Using the expansion, and keeping just the first term

$$(1.01)^{10} = (1 + 0.01)^{10}$$

$$= 1 + 10 \times 0.01$$

$$= 1.10$$

This differs by less than one-half of one per cent from the correct value. Adding the next term in the expansion would give

$$(1.01)^{10} = 1 + 10 \times 0.01 + 10 \times 9 \times (0.01)^2 / 2$$

$$= 1.1045$$

This result differs by only 0.01 per cent from the actual value.

One must be careful in using such series approximations. For example, evaluate $(1.05)^{10}$. This is actually 1.6289, but using the expansion to one term,

$$(1 + 0.05)^{10} = 1 + 10 \times 0.05$$

$$= 1.5$$

The error is large, 8 per cent. Keeping one more term

$$(1 + 0.05)^{10} = 1 + 10 \times 0.05 + 10 \times 9 \times (0.05)^2 / 2$$

$$= 1.6125$$

The error is down to 1 per cent.

The expansion can be used for more than numerical work. An example in a theoretical development follows. The mass of an object on the basis of special relativity theory is given by

$$m = \frac{m_o}{\sqrt{1 - v^2/c^2}}$$

where m_o is the rest mass, c is the speed of light, and v is the speed of the object. Assume that $v \ll c$, and write it in the form

$$m = m_o \ (1 - v^2/c^2)^{-\frac{1}{2}}$$

$$\text{apply } (1 + x)^p = 1 = px$$

$$\text{where } x = -v^2/c^2 \text{ and } p = -\frac{1}{2}$$

$$\text{then } m = m_o \ (1 + \frac{v^2}{2c^2})$$

$$\text{or } m = m_o + \frac{1}{2} \frac{m_o v^2}{c^2}$$

The last term is the increase in mass; let it be Δm, and

$$\Delta m = \frac{1}{2} \frac{m_o v^2}{c^2}$$

This expression can be used to evaluate very simply a mass increase. The fractional increase, $\Delta m/m_o$, is just $v^2/2c^2$. It is obvious that the speed v must be very high for this to be significant.

While we are at this point, consider two things: the common expression for kinetic energy, $m_o v^2/2$, and the relation that is associated with Einstein and relativity, that $E = mc^2$. Multiplying both sides of $\Delta m = m_o v^2/2c^2$ by c^2 yields

$$\Delta mc^2 = \frac{1}{2} m_o v^2$$

The left side is the energy associated with a mass moving at a speed v, and the right side is the familiar expression for kinetic energy. At low speeds, those for which $v/c \ll 1$, relativity theory gives the same expression for energy of motion as does classical theory, and this has been demonstrated using the binomial expansion.

2

APPENDIX

NATURAL LOGARITHMS

In the table below are the natural logarithms for numbers between 1 and 10. These are logarithms to the base e and are denoted by the abbreviations $\log x$ or $\ln x$. They are the power to which e $(2.718\ldots)$ is raised to obtain the given number.

x	.0	.1	.2	.3	.4	.5	.6	.7	.8	.9
1	0.00	0.10	0.18	0.26	0.34	0.41	0.47	0.53	0.59	0.64
2	0.69	0.74	0.79	0.83	0.88	0.92	0.96	0.99	1.03	1.06
3	1.10	1.13	1.16	1.19	1.22	1.25	1.28	1.31	1.34	1.36
4	1.39	1.41	1.44	1.46	1.48	1.50	1.53	1.55	1.57	1.59
5	1.61	1.63	1.65	1.67	1.69	1.70	1.72	1.74	1.76	1.77
6	1.79	1.80	1.82	1.84	1.86	1.87	1.89	1.90	1.92	1.93
7	1.95	1.96	1.97	1.99	2.00	2.01	2.03	2.04	2.05	2.07
8	2.08	2.09	2.10	2.12	2.13	2.14	2.15	2.16	2.17	2.19
9	2.20	2.21	2.22	2.23	2.24	2.25	2.26	2.27	2.28	2.29
10	2.30									

To find the natural logarithm of a number which is outside the range of the table, express it as a number between 1 and 10 times some power of 10. Find in the table the logarithm of the number now between 1 and 10 and add to it $n \log_e 10$, where n is the power of 10 involved. If the exponent is negative, subtract $n \log_e 10$. Some values of $n \log_e 10$ are listed below.

n	$n \ \log_e \ 10$
1	2.30
2	4.61
3	6.91
4	9.21
5	11.51
6	13.82

Examples

1. To find the natural logarithm of 3400, write it as 3.4×10^3.
    The $\log_e 3.4 = 1.22$
    And $\log_e 10^3 = 3 \times \log_e 10 = 6.91$
    Added together they give 8.13
    That is: $\log_e 3400 = 8.13$
2. To find the natural logarithm of 0.00059, write it as 5.9×10^{-4}.
    The $\log_e 5.9 = 1.77$
    And $4 \times \log_e 10 = 9.21$
    Subtract to get $1.77 - 9.21 = -7.44$,
    That is: $\log_e 0.00059 = -7.44$

INDEX